BETTER BY DESIGN

HOW USER EXPERIENCE ACCELERATES GROWTH, LOYALTY & INNOVATION

MIKE KUECHENMEISTER

Better by Design: How User Experience
Accelerates Growth, Loyalty, and Innovation
1st Edition

Introduction

Introduction

Why do some companies consistently achieve extraordinary results, driving customer loyalty, operational efficiency, and revenue growth, while others struggle to keep up?

The answer often lies in a discipline many leaders overlook: User Experience (UX). For companies like Capital One, Sephora, and Southwest Airlines, UX is not just a design practice. It's a core business capability. By embedding UX into their organizations, these companies have created seamless, intuitive experiences that delight customers, empower employees, and deliver measurable results.

The data is compelling: research shows businesses investing in UX see an average ROI of 9,900%—$100 in returns for every $1 spent. More on that later.

UX doesn't have to be a mystery. It's not an abstract concept reserved for tech startups. UX is a practical, proven approach to solving real business problems. I wrote this book to demystify UX and show you how to unlock its value for your organization.

My Journey to UX

I began my career in advertising with a clear goal: to become a creative director by the age of 28. By 27, I had achieved that milestone, leading creative efforts at one of the top agencies in the Southeast at that time. Eventually, I left the agency world to start my practice, seeking creative freedom and balance.

In those early days, I made plenty of mistakes, including launching a startup without the benefit of modern UX practices like prototyping and user testing. This experience was humbling and invaluable, teaching me how essential it is to deeply understand your users and iterate to create better solutions. My career has taken me across the worlds of healthcare, finance, and non-profits.

Eventually, I joined a Fortune 100 company to help lead design. There, I scaled high-performing UX teams and aligned UX with digital transformation efforts. I've worked with startups and enterprises alike, seeing what drives success, and what leads to failure.

Throughout my journey, I've remained a student of business, fascinated by the intersection of strategy, creativity, and execution. This book is my way of sharing what I've learned, bridging those worlds, and showing leaders how UX can be a catalyst for sustainable growth and innovation.

Why I Wrote This Book

Too often, UX is misunderstood, underutilized, or treated as an afterthought. It's time to change that. UX can drive extraordinary value for any organization. But to realize that value, it must be understood and embraced. My goal with this book is simple: to give you a clear, actionable understanding of UX. You'll learn what UX is, how it works, and why it's one of your business's smartest investments.

This book is about more than understanding UX. It's about unlocking its potential to create meaningful and measurable change. When you finish, you'll understand UX's extraordinary ROI and have the tools and knowledge to unlock it within your organization.

I've been on this journey—making mistakes, learning lessons, and experiencing firsthand what works. Now, I want to share those insights with you. Let me guide you as we explore how UX can elevate your business, delight your customers, and empower your teams.

Chapter 1:
Driving Growth Through UX

The goal is to provide users with the best possible user experience in a way that provides your organization with the best possible business results.

Jeff Horvath
Vice President UX Strategy
Human Factors International

Does this sound familiar?

Your operations rely on a tangled web of disconnected tools, complicating what should be straightforward processes. Your sales team struggles with CRM systems that don't integrate seamlessly, delaying deal cycles and creating frustration. Customers encounter friction at critical moments—whether through confusing checkout experiences, inconsistent support, or communication gaps across channels. Employees lose hours navigating outdated internal systems, which drains productivity.

Most concerning of all, you're unsure which features of your products or services deliver real value, leaving you guessing where to focus your investments.

For many businesses, these challenges are natural byproducts of growth. Over time, systems, tools, and workflows are added to address immediate needs. But this layering creates complexity, resulting in a patchwork of solutions that aren't built to scale. The consequences are clear: bottlenecks, inefficiencies, and missed opportunities ripple through your operations, impacting strategy, sales, customer support, and overall performance. If this resonates, you're not alone. These are the kinds of problems UX is uniquely equipped to solve.

UX is about creating systems that meet user needs while driving business objectives

Don Norman, who first coined the term "user experience," defines it as encompassing "all aspects of the end-user's interaction with the company, its services, and its products." This definition emphasizes that UX is far more than surface design or isolated touchpoints. It's about creating systems that meet user needs while driving business objectives. Effective UX identifies pain points, reduces complexity, and aligns workflows with how people interact with tools and services.

By addressing issues such as confusing interfaces or inefficient processes, UX enables customers to complete tasks seamlessly, employees to work effectively, and businesses to achieve their strategic goals.

In today's competitive landscape, companies that prioritize seamless, intuitive experiences stand out. UX transforms inefficiencies and frustrations into opportunities for operational efficiency, trust-building, and user satisfaction.

The Role of UX in Driving Results

Think of UX as the infrastructure that enables people to achieve their goals quickly and effortlessly. When executed well, UX removes friction, saves time, and leaves a positive impression, ultimately driving better outcomes for your business. It's not about appearances but about functionality and usability.

Although UX is often associated with digital interfaces such as apps and websites, its impact extends far beyond screens. Every interaction a user has with your product, tool, or service is part of the user experience. This could be a customer navigating your website, an employee using an internal dashboard, or a client filling out a service form. These interactions, large or small, shape perceptions and influence outcomes.

Distinguishing UX from CX

Understanding the distinction between UX and Customer Experience (CX) is crucial. CX encompasses the entirety of a customer's journey with your brand, including every touchpoint and emotional connection. UX, in contrast, focuses on the usability and functionality of specific interactions.

For example, CX evaluates how a customer feels about your brand after a support interaction. UX ensures that the support portal itself is intuitive, fast, and effective at resolving issues. While both disciplines are critical, UX often serves as the foundation for delivering exceptional CX. Without smooth, intuitive interactions, even the best CX strategy can fall short.

Why UX Is Essential for Business Success

When UX is effective, it creates seamless experiences that reduce friction, improve efficiency, and drive satisfaction. Customers complete tasks effortlessly, employees work more efficiently, and processes run more smoothly. Though often invisible to the end user, UX plays a vital role in the success of businesses across industries. It's not just about creating functional systems; it's about delivering satisfying, impactful experiences that align with your business goals.

By prioritizing UX, you're not merely addressing pain points, you're unlocking the potential for sustained growth, innovation, and excellence throughout your organization.

UX Is Everywhere

UX is already a part of your business, whether intentionally designed or not. Every interaction someone has with your website, product, service, or internal tools reflects decisions made by your team, deliberate or otherwise. These interactions collectively define the user experience. When they fall short, the consequences ripple through your organization, affecting satisfaction, efficiency, and revenue.

For example, consider a website that makes it difficult for customers to find what they need. That frustration is a reflection of poor UX. Employees wasting hours navigating disjointed systems to complete tasks are victims of suboptimal UX. Sales teams forced to work with a CRM that requires manual data entry experience UX flaws that delay deals and create missed opportunities. If your onboarding process leaves new customers unsure of how to use your product, that's another example of ineffective UX. The same applies to clunky ticketing systems that slow support teams, overwhelming interfaces that cause users to abandon your product, or inconsistencies between mobile and desktop experiences that confuse users. Every one of these interactions shapes perceptions of your business.

Good UX fosters clarity, ease, and satisfaction.

Poor UX breeds friction, confusion, and inefficiency. Recognizing UX as a foundational element of your business is the first step toward taking control. By prioritizing user experience, you set the stage for stronger relationships with your customers, higher productivity for your employees, and sustained growth for your business.

The Many, Many Names of UX

UX is a multifaceted discipline and has gone by many names over the year. The different names reflect various perspectives and areas of expertise within the field as it has evolved. These terms are sometimes used interchangeably, but each brings its own nuance, illustrating the depth and evolution of

UX as a practice. For this book, we take a broad approach, focusing on UX's ability to create seamless, effective, and impactful interactions across every touchpoint. Within your own organization, you might recognize some of these common names for UX:

Product Design focuses on creating tangible and digital products emphasizing functionality, usability, and aesthetics. While it often overlaps with UX, product design concentrates on the specific features, workflows, and interfaces of individual products. It's about crafting solutions that meet user needs while maintaining visual and functional coherence within a single offering.

Human-centered design (HCD) is a design philosophy that centers problem-solving on human needs, behaviors, and limitations. Broader than UX, HCD applies to various disciplines, from industrial design to systems engineering. It emphasizes empathy and iterative processes to create solutions that resonate with users and address their real-world challenges.

User Interface (UI) Design focuses on the visual and interactive elements of digital interfaces, including buttons, menus, layouts, and typography. As a subset of UX, UI design is specifically concerned with the aesthetics and usability of digital touchpoints, ensuring that they are visually appealing and easy to navigate. However, it does not address the broader user journey or system functionality.

Human Factors is the study of how humans interact with systems, products, and environments, often in high-stakes industries like healthcare or aviation. This discipline prioritizes ergonomics, safety, and efficiency using a scientific and engineering-based approach. Human factors aim to optimize interactions to ensure usability and minimize errors, especially in critical settings.

Service Design encompasses the entire service journey, including customer-facing interactions and behind-the-scenes processes. While it overlaps with UX, service design includes operational elements such as staffing, logistics, and service delivery systems. Its goal is to create seamless, end-to-end experiences that align with user needs and business objectives.

Interaction Design (IxD) focuses on how users engage with and navigate a product or system, emphasizing behaviors, flows, and feedback mechanisms. As a narrower subset of UX, interaction design hones in on the dynamic elements of user interactions, ensuring that they are intuitive, efficient, and satisfying. It plays a critical role in creating seamless and engaging user experiences.

Each of these terms represents a piece of the UX story, highlighting different aspects of creating and improving experiences. But whether you're optimizing a single feature, rethinking an entire service, or aligning internal workflows with user needs, the principles of UX remain the same.

Key Functions Within UX

The structure of UX teams can vary significantly from one company to the next. Titles, specialties, and organizational models often differ across industries and business sizes. But regardless of these variations, UX consistently relies on four core functions: research, design, writing, and strategy. These disciplines work together to ensure user needs align seamlessly with business objectives, resulting in intuitive, impactful experiences that align with organizational goals.

Research is where it all begins. The goal is to deeply understand user behaviors, needs, and pain points to inform design decisions. UX researchers gather insights through interviews, surveys, usability tests, and data analysis. These insights are distilled into personas, journey maps, and usability reports that clearly depict user goals and challenges.

For example, research might uncover why customers abandon a checkout process or what employees find frustrating about internal tools. Addressing these real-world problems ensures that UX efforts are grounded in evidence and targeted at delivering meaningful results.

Design is the process of translating research into intuitive and visually engaging solutions. UX designers combine interaction design, visual design, and information architecture to craft experiences that guide users seamlessly through interactions.

Designers create wireframes, prototypes, and high-fidelity designs while maintaining consistency through design systems and style guides. Whether simplifying workflows for internal teams or enhancing brand perception through polished interfaces, the design function transforms user insights into tangible experiences that solve problems and delight users.

Content plays a critical role in the user experience, and UX writers ensure that every word contributes to clarity and consistency. Microcopy, such as button labels, error messages, and instructional text, guides users through their journey while reflecting the brand's tone and voice.

By collaborating closely with designers and researchers, writers reduce confusion and create a human connection between the user and the interface. Thoughtful writing doesn't just support the design, it elevates it, ensuring that users understand and trust every step of the experience.

Why These Functions Matter

Together, these four functions form the backbone of a successful UX practice. Researchers uncover the insights that drive the process. Designers turn those insights into solutions. Writers ensure that the experience is straightforward and empathetic. Strategists align everything with business priorities. When these functions operate in harmony, UX becomes a driving force for innovation, growth, and user satisfaction.

UX for External and Internal Users: Two Sides of the Same Coin

Great customer experiences and efficient employee tools are inseparable. Customer interactions with your business, whether through your website, app, or support channels, depend heavily on the systems and tools your employees use behind the scenes. When internal systems are intuitive and streamlined, employees can focus on delivering exceptional service. Similarly, smooth customer-facing experiences reduce the strain on your teams, creating a virtuous cycle: happy, empowered employees deliver outstanding experiences to satisfied, loyal customers. We'll unpack this concept as we go along.

UX for External Users

Customer-facing UX is often the most visible part of the equation, where businesses directly engage with their audiences. We'll double click on this in Chapter 4. Great UX builds loyalty, encourages repeat business, and turns satisfied customers into advocates for your brand. What defines great UX? It's about removing obstacles, simplifying processes, and creating experiences that feel effortless.

Consider these examples:

- **Retail**: An intuitive website allows customers to find, compare, and purchase products quickly and easily without confusion or delays.

- **Hospitality**: A hotel check-in process, whether through self-service kiosks or staff, flows so smoothly that guests never feel frustrated or inconvenienced.

- **Food Service**: A restaurant chain's mobile app allows customers to customize meals, pay ahead, and pick up their orders seamlessly, avoiding lines.

- **Banking**: An online platform enables customers to open accounts in minutes with clear guidance and instant confirmation, removing the need for branch visits.

- **Travel**: An airline app provides real-time updates, gate information, and rebooking options in just a few taps, easing travel disruptions.

- **Fitness**: A subscription service app personalizes workout plans, remembers preferences, and syncs progress across devices, motivating users to stay engaged.

- **E-commerce**: An online marketplace suggests optimized shipping options based on location and needs, making purchasing effortless.

Take Starbucks, for example. Starbucks isn't just about coffee, it's about the experience. The company's mobile app allows customers to skip lines. At the same time, its store layouts are designed for comfort and efficiency. Every interaction, whether digital or in-store, is crafted to reduce friction and enhance convenience. Starbucks delivers not just a product but a consistently satisfying experience, encouraging customers to return time and time again. Even more, with coffee as their core product, they make 36 billion a year and enjoy a cult-like following.

Meeting Rising Consumer Expectations

Consumer expectations have never been higher. Technological advances and seamless experiences provided by industry leaders have redefined what customers expect: intuitive, personalized, and connected interactions across every touchpoint. Frictionless digital experiences are now the norm, matching the convenience and responsiveness of leading apps and platforms.

Businesses that fail to meet these expectations risk losing customers to competitors who prioritize UX and deliver thoughtful, seamless designs. Investing in customer-facing UX is no longer optional. It's essential for building loyalty and remaining relevant in a market where user-centric design sets the standard for success.

The Interdependence of UX for Employees and Customers

Customer experiences are only as strong as the systems supporting them. By ensuring that internal tools are as thoughtfully designed as customer-facing ones, businesses create a foundation for success. Investing in both sides of UX, external and internal, empowers employees, delights customers, and drives business outcomes, proving that these two sides are indeed one cohesive system.

UX for Internal Users

We'll upack this in Chapter 5, but in short, Internal UX drives efficiency, boosts employee satisfaction, and enhances overall organizational performance. Thoughtfully designed systems that streamline workflows and eliminate unnecessary friction empower employees to focus on high-impact tasks and achieve better results. Whether simplifying routine processes or enabling large-scale digital transformation, effective internal UX helps employees work smarter and more effectively. Here are some examples of where internal UX can make a difference:

Human Resources
A redesigned benefits portal makes it easy for employees to enroll in plans and track their options. This improvement reduces HR inquiries, speeds decision-making, and enhances employee satisfaction.

Centralized Dashboards for Digital Transformation
A multinational organization develops a unified dashboard combining payroll systems, benefits portals, productivity apps, and third-party tools like Google Workspace. With everything accessible from a single interface, employees save time and boost productivity.

Sales Teams
A CRM system streamlined with a cleaner interface and simpler workflows reduces the time sales reps spend on data entry. This shift allows them to focus on closing deals. Research shows that sales teams using optimized tools are significantly more likely to meet their revenue targets.

Customer Support Command Centers
A tech company integrates customer communication channels, ticket management, and third-party analytics into a centralized support platform. Agents resolve issues faster and deliver personalized service, improving employee satisfaction and customer outcomes.

IT Helpdesks
Automating common requests, such as password resets, through a user-friendly self-service portal reduces IT workload and response times. Employees spend less time waiting for support and more time focusing on their tasks.

Procurement Systems
A company consolidates multiple procurement platforms into a single interface with automated approval workflows. Employees complete purchasing tasks quickly and efficiently, ensuring compliance and saving time.

Field Operations

A utility company equips field technicians with a mobile app
that integrates job updates, route optimization, inventory
tracking, and diagnostic tools. Technicians can access
everything they need on the go, improving service delivery and
reducing downtime.

Training and Onboarding

A logistics company develops an interactive onboarding
platform with gamified elements, helping new hires learn
processes more effectively. This reduces ramp-up times and
lowers turnover rates among new employees.

Finance Teams

A finance department revamps its expense reporting system
with features like mobile receipt scanning and pre-filled
data fields. These improvements reduce errors, shorten
reimbursement times, and save time for employees and finance
staff.

Internal UX and Digital Transformation

Internal UX eliminates redundancies and creates cohesive,
intuitive systems by consolidating tools, streamlining
workflows, and integrating third-party solutions. These
enhancements make employees more productive, improve
morale, and reduce turnover, all while improving the quality
of service delivered to customers. In the context of digital
transformation, internal UX becomes a key driver of innovation
and operational excellence. It enables employees to thrive in an
increasingly connected workplace, helping organizations adapt
to new challenges and unlock their full potential.

As a global technology leader, IBM encountered significant issues with its outdated and fragmented internal systems. These legacy tools and inefficient workflows not only decreased employee productivity but also impacted morale, ultimately affecting the company's ability to deliver exceptional customer service.

Implementing Enterprise Design Thinking

Recognizing the need for a systemic change, IBM leveraged its Enterprise Design Thinking framework to address these inefficiencies. This approach focused on enhancing user-centered design, fostering collaboration, and applying iterative problem-solving techniques to overhaul its internal systems. The initiative began with extensive user research, where IBM's design team conducted interviews, surveys, and usability testing to pinpoint critical pain points, such as cumbersome navigation and disjointed workflows.[1]

Prototyping and Iterative Development

With actionable insights from the research phase, IBM's team developed prototypes for new tools and workflows. These were continuously tested and refined with direct employee feedback, ensuring the solutions were practical and met the real-world needs of the workforce. The incremental rollout of these redesigned tools, paired with comprehensive training and support, facilitated a smooth transition for employees, fostering widespread adoption and minimal disruption.

Measurable Outcomes and Strategic Impact

The revamp of IBM's internal systems yielded substantial benefits. According to a Forrester study, the project delivered a 301% ROI over three years. Moreover, streamlined workflows and integrated tools saved employees over 275,000 hours annually, allowing them to focus on higher-value tasks such as innovation and strategic planning. The enhancements in system usability boosted productivity, reduced delays and errors, and significantly improved employee satisfaction and engagement.

The improvements in internal UX extended beyond IBM's immediate workforce. Empowered employees could address customer needs more effectively, leading to faster issue resolution and higher client satisfaction. The success of this initiative not only reinforced IBM's reputation as a leader in innovation and user-centered design but also demonstrated the profound impact of internal UX on overall business performance.

Why UX Matters More Than Ever

We're operating in a world where expectations, both from customers and employees, are higher than ever. Globalization, rapid technological advancements, and shifting geopolitical landscapes have redefined how businesses must perform to stay competitive. Customers now demand seamless, personalized interactions that anticipate their needs, while employees expect tools that simplify workflows and enable them to focus on meaningful, high-value tasks. Generic, one-size-fits-all solutions are no longer enough. Companies that fail to meet these rising expectations risk losing their most loyal customers and top talent to competitors who prioritize exceptional experiences.

This is where UX comes in. When UX is treated as a core business function, it equips organizations to adapt, innovate, and thrive continuously in this rapidly changing environment.

UX empowers businesses to thrive in an ever-changing world by delivering resilient and adaptive solutions. It creates reliable experiences that maintain continuity and consistency, even amid external disruptions. By harnessing AI, machine learning, blockchain, and automation advancements, UX enables more innovative, more efficient interactions that meet users' evolving demands. Personalization becomes a competitive advantage as UX designs tools and touchpoints tailored to customers' and employees' unique preferences and behaviors.

UX empowers businesses to thrive in an ever-changing world by delivering resilient and adaptive solutions.

Internally, UX improves employee productivity by streamlining tools and workflows, reducing frustration, and enabling teams to focus on high-impact tasks. Externally, it strengthens customer retention by building trust and loyalty through seamless and consistent experiences that exceed expectations. UX also accelerates innovation, leveraging iterative design and rapid prototyping to refine ideas quickly and maintain a competitive edge.

UX's influence extends to market expansion. It adapts experiences for global audiences, addressing cultural, linguistic, and regulatory nuances without compromising quality. It enhances accessibility, ensuring products and services are inclusive and usable for a broader audience, reflecting a commitment to equity. Behavioral insights, feedback, and analytics drive continuous optimization of customer and employee experiences, enabling businesses to make data-driven decisions that improve outcomes.

Enhancing Operational Efficiency

Finally, UX contributes to operational efficiency by eliminating redundancies, streamlining processes, and saving time and resources across the organization. Whether it's increasing retention, reducing costs, or enabling innovation, UX provides the tools and strategies businesses need to succeed in an increasingly dynamic environment.

You lay the groundwork for sustainable growth and adaptability by embedding UX into your business strategy. It equips your organization to meet rising customer and employee expectations and capitalize on new opportunities in an ever-changing market. However, unlocking the full potential of UX requires more than isolated efforts. It demands alignment with your overarching business goals. UX must become a measurable contributor to your company's success. In the next chapter, we'll dive into the ROI of UX, exploring its tangible value and why it's one of the smartest investments your business can make.

NOTES

NOTES

Chapter 2:
The ROI of UX

Good design is good business.

Thomas J. Watson Jr.
former CEO—IBM

Investing in UX delivers measurable financial returns and transformative business outcomes. According to Forrester[2], companies prioritizing user-centered design see an average ROI of 9,900%, meaning every dollar invested in UX can generate $100 in value. This remarkable return underscores the power of UX to address critical business challenges holistically while driving growth, loyalty, and efficiency.

McKinsey & Company's research[3] further validates the business case for UX. Their study on the Business Value of Design found that companies with top-quartile McKinsey Design Index scores outperformed industry benchmarks by nearly two-to-one in revenue growth and generated shareholder returns that were significantly higher than their peers. These companies excel by integrating design principles into their strategic decision-making and day-to-day operations, ensuring that UX contributes to user satisfaction and business performance.

Reducing Friction, Empowering Users, and Mitigating Risks

At the core of UX's extraordinary ROI is its ability to reduce friction, empower users, and mitigate risk. Streamlined workflows and intuitive interfaces eliminate inefficiencies, enabling users to achieve their goals with less effort. This reduction in friction directly impacts key revenue drivers, such as increased conversion rates, higher subscription renewals, and greater customer lifetime value. Empowered users, whether customers or employees, experience higher satisfaction and trust, while intuitive systems boost employee engagement and productivity, enabling teams to focus on innovation and strategic goals. Additionally, UX mitigates risk by addressing usability issues early in the design process. By uncovering and resolving potential problems before launch, businesses reduce costly post-launch corrections, improve product success rates, and ensure alignment with user needs and business objectives.

To quantify these outcomes, frameworks like Forrester's ROI analysis and McKinsey's Design Index provide clear metrics. Revenue metrics, such as sales growth and customer retention rates, demonstrate the financial impact of improved UX. Operational savings from reduced support inquiries and workflow inefficiencies highlight cost benefits. Enhanced employee productivity metrics, including shorter task completion times and higher job satisfaction, showcase the internal advantages of user-centered design. Together, these metrics offer a comprehensive roadmap for understanding the tangible value of UX.

By addressing inefficiencies, empowering users, and reducing risks, UX enables businesses to achieve sustainable growth, outperform competitors, and deliver measurable returns for stakeholders.

Real-World Examples of UX ROI

Forrester and McKinsey's findings are not just numbers. They reflect the real-world success of companies that have leveraged UX strategically:

Redefining Mobile Banking

Bank of America recognized the critical role of its mobile app as a primary touchpoint for customers. The bank launched a significant update to enhance engagement, consolidating five separate apps into a unified platform.[4] This redesign emphasized ease of use, personalized financial insights, and seamless integration of services, enabling users to manage their finances more effectively. The results were notable. The bank's digital clients, totaling 57 million, benefited from a more personalized digital experience, leading to increased digital engagement. The previous year, clients connected with their finances digitally a record 23.4 billion times, marking an 11% increase year-over-year. This surge in digital interactions reflects improved usability and customer trust. Additionally, the unified app received the 2024 Celent Model Bank award for customer-centered innovation, underscoring its success in meeting user needs.

Optimizing Through A/B Testing

Booking.com exemplifies the power of continuous A/B testing in enhancing user experience and driving business success. By persistently experimenting with elements such as button placements, color schemes, and navigation flows, the company refines its platform to align with user expectations.[5] This culture of relentless iteration has contributed to Booking. com achieving one of the highest conversion rates in the travel industry, significantly boosting annual revenue. Users benefit from seamless, frustration-free interactions, enhancing customer satisfaction and loyalty.

Increasing Revenue Through Checkout Optimization

Target identified a significant drop-off during its online checkout process, prompting the company to streamline the experience by reducing the number of steps required to complete a purchase and introducing features like auto-filled payment information.[6] While specific figures are proprietary, such optimization efforts in e-commerce have been shown to substantially increase completed transactions, generating millions in additional annual revenue. Customers experience less friction during checkout, improving satisfaction and reducing cart abandonment.

Transforming Air Travel Through Digital Experiences

Delta Air Lines significantly enhanced its Fly Delta app to address common customer pain points, such as last-minute gate changes and flight delays. The redesigned app offers

real-time updates, easy rebooking options, and baggage
tracking, creating a seamless digital travel companion.[7] These
improvements have increased customer satisfaction and Net
Promoter Scores (NPS). According to Comparably, Delta's
NPS is 43, with 64% of customers identified as Promoters,
indicating a strong likelihood to recommend the airline.
Travelers experience reduced stress during their journeys
as the app transforms traditionally frustrating moments into
smooth, well-informed experiences, fostering increased loyalty
and repeat bookings. Delta's commitment to enhancing the
customer experience through digital innovation is evident in
the app's continuous updates, including the recent Fly Delta 6.0
release, which introduced features like an all-new Help Center
view and improved access to boarding information. We'll come
back to Delta's app redesign in Chapter 4.

Streamlining Employee Scheduling

Kaiser Permanente recognized the need to modernize its
employee scheduling system, which had been used for three
decades. The organization implemented a new workforce
management technology to replace the outdated DOS-based
system, aiming to enhance operational efficiency and reduce
scheduling errors.[8] While specific metrics are proprietary,
such system overhauls in healthcare settings have significantly
reduced scheduling errors, leading to improved operational
efficiency. Additionally, such improvements increase
employee satisfaction, as staff experience less frustration with
administrative tasks, allowing them to focus more on patient
care. Enhanced scheduling systems also support better patient
care by ensuring appropriate staffing levels, reducing employee

stress, and enabling healthcare workers to deliver higher-quality service.

From User Success to Business Impact

To fully leverage UX as a strategic advantage, it's essential to understand the role of user success metrics and how they connect to broader business KPIs. User success metrics measure how effectively users achieve their goals when interacting with your product, service, or system. These metrics are often leading indicators of business performance, offering early insights into future outcomes like revenue growth, customer retention, and operational efficiency.

User success metrics focus on the quality and effectiveness of user interactions. They provide a detailed picture of how well your systems, tools, or experiences meet users' needs. Some common examples include:

Task Completion Rate: The percentage of users who successfully complete a specific action, such as booking a flight or completing a purchase.

Time on Task: How long it takes users to complete an interaction, such as filling out a form or navigating through a dashboard.

Error Rate: The frequency with which users encounter issues, such as filling out incorrect fields in a form or making a mistake due to unclear instructions.

User Satisfaction (SUS or CSAT) Measures how satisfied users are with a specific experience. It is often captured through surveys or usability tests.

Drop-off Rate: The percentage of users who abandon an action midway, such as leaving a checkout process before completing the purchase.

Adoption Metrics: The rate at which users engage with new features or tools, indicating how effectively these innovations meet user needs.

UX metrics are crucial drivers of business success, serving as early indicators of broader performance trends. For example, simplifying the checkout process can dramatically increase conversion rates. Research shows that large e-commerce sites can boost conversions by over 35% through optimized checkout design, directly contributing to revenue growth.[9] Similarly, according to statistics from Landingi, improving UX design can increase conversions by up to 400%, illustrating the profound impact of user-centered design on revenue.[10]

UX is a strategic approach to unlocking sustainable growth and competitive advantage.

High satisfaction scores (CSAT) and low error rates during onboarding also have measurable benefits for customer retention. A SaaS company that redesigned its onboarding process to address user pain points saw churn drop by 20%, boosting customer lifetime value. Effective UX/UI design has been shown to enhance conversion rates by up to 200%, guiding users toward meaningful interactions and reinforcing their trust in the brand.[11]

Operational efficiency is another area where UX delivers tangible results. Forrester research highlights that optimizing internal tools can save 20–30% of the time across workflows.[12] Streamlining internal systems reduces task complexity, lowers operational costs, and improves employee satisfaction and productivity. These improvements empower teams to focus on strategic, high-value tasks, further enhancing business performance.

Focusing on metrics like task completion rates, error reductions, and user satisfaction can help businesses achieve measurable improvements in revenue, retention, and efficiency. These examples illustrate that investing in UX is not just about improving user experiences. It's a strategic approach to unlocking sustainable growth and competitive advantage.

Aligning UX Metrics with Business KPIs

To maximize the impact of UX, it's critical to align user success metrics with broader business goals. Companies can ensure that UX initiatives drive meaningful and measurable results by connecting specific UX outcomes to organizational priorities. Here are several examples of how this alignment works across different goals and contexts:

Customer Retention

Retention is one of the most cost-effective ways to drive business growth, as keeping existing customers is significantly less expensive than acquiring new ones. UX is critical in fostering customer loyalty by enhancing the touchpoints that matter most. Metrics such as onboarding task completion rates are essential indicators of success. User satisfaction, often measured through CSAT surveys, provides additional insight into how users perceive their early experiences and subsequent interactions with a product or service. Feature adoption metrics, which track whether customers engage with core functionalities, serve as another valuable measure, signaling long-term engagement and the likelihood of continued retention. Together, these metrics demonstrate how a strong UX foundation can turn first-time users into loyal customers.

Operational Efficiency

Operational efficiency is a cornerstone of scaling businesses while maintaining control over costs, and UX improvements play a vital role in achieving this. By optimizing internal systems, companies can streamline workflows, reduce errors, and save employees valuable time. Metrics such as employee task completion times provide clear insight into efficiency gains. Error rates are another critical measure, as poorly designed tools often lead to mistakes and costly rework. Reducing these errors not only improves accuracy but also boosts overall productivity. Additionally, tool utilization metrics can reveal whether employees fully leverage available systems or rely on inefficient workarounds, signaling areas for improvement. These metrics highlight how targeted UX enhancements can drive operational efficiency and unlock significant value for organizations.

Driving Revenue Growth

Revenue growth depends on removing friction from customer interactions and optimizing user journeys to encourage conversions. UX is critical in streamlining touchpoints and ensuring users can accomplish their goals effortlessly. Metrics such as drop-off rates provide valuable insights into areas where users abandon processes like checkout. Conversion rates, which measure the percentage of users completing desired actions, such as purchasing or signing up for a service, are another essential indicator of success. Task completion times highlight how quickly customers can achieve their goals, whether booking a flight or setting up an account. Faster interactions often lead to higher satisfaction, increased loyalty, and more sales, demonstrating how UX improvements directly contribute to revenue growth.

Enhancing Customer Satisfaction and Brand Loyalty

Customer satisfaction and loyalty are essential for building a strong brand and creating advocates who actively promote your business. UX improvements at critical touchpoints, such as customer support, self-service portals, or digital interfaces, significantly influence these outcomes. Metrics like Net Promoter Score (NPS) provide valuable insights into how likely customers are to recommend your brand, reflecting overall satisfaction and loyalty.

Measuring satisfaction with digital interactions, including online tools like chatbots or self-service platforms, helps identify opportunities to enhance user experiences. Addressing churn rate, by resolving customer journey friction during onboarding or billing, further strengthens retention and ensures long-term engagement. Businesses can create seamless, satisfying experiences that foster loyalty and build advocacy by focusing on these key areas.

Supporting Employee Engagement and Retention

Employee engagement and retention are critical for operational success, as engaged employees are more productive and less likely to leave. Enhancing internal UX directly supports these outcomes by providing tools and systems that are intuitive and efficient, reducing frustration and enabling employees to focus on high-impact tasks.

Metrics such as internal satisfaction scores, gathered through surveys, offer insight into how employees feel about the tools they rely on daily. Measuring time-to-competency tracks how quickly new employees adapt to internal systems, with shorter onboarding periods indicating more effective and user-friendly tools. Turnover rates also reflect the impact of internal UX, as well-designed systems reduce frustration and contribute to a more engaged, stable workforce. By prioritizing internal UX, organizations create a positive environment that drives productivity and minimizes the costs associated with employee attrition.

Enabling Innovation and Agility

Prioritizing UX enables organizations to innovate and adapt more effectively, allowing them to respond quickly to market changes and evolving user needs. By embedding UX into their processes, companies can foster a culture of continuous improvement that drives both innovation and agility. Metrics like feature feedback and iteration rates highlight how efficiently teams can collect user input and refine their solutions. Prototype validation metrics, gathered through usability testing, ensure that new ideas are tested and optimized before full development. This approach minimizes risks and maximizes the chances of delivering successful outcomes. By leveraging these metrics, organizations can stay ahead of the competition, continuously enhancing their offerings while aligning with user expectations.

Leading vs. Lagging Indicators

Traditional business KPIs like profit margins, customer retention rates, and employee turnover are invaluable for assessing organizational performance but often reflect the results of decisions made months or even years earlier. UX success metrics, by contrast, provide a forward-looking lens. These leading indicators focus on user behaviors, attitudes, and interactions that drive broader business outcomes, offering early insights into what's working and where course corrections are needed. This allows leadership to act proactively, optimizing performance before lagging metrics like revenue or churn rates fully materialize.

Lagging Indicators	Leading Indicators
Revenue Growth	Decreased task completion time leading to increased transactions
Employee Productivity and Operational Costs	Higher usability scores in internal tools
Customer Retention and Lifetime Value	Increased customer satisfaction during onboarding
User Engagement and Retention	Reduced drop-off rates in product usage
Customer Support Costs and Customer Satisfaction	Increased usage of self-service portals
Product Adoption and Success	Positive feedback in prototype usability testing
Market Share	Early adoption rates of new features or products
Brand Loyalty	Improved Net Promoter Scores (NPS) following UX improvements
Return on Investment (ROI)	Increased efficiency in user workflows
Customer Acquisition Costs	Enhanced usability reducing the need for extensive customer support
Quality of Service Delivery	Lower error rates in user interactions
Operational Agility	Quicker adaptation to user feedback in product updates

Consider an e-commerce site that reduces task completion time, a UX metric that reliably predicts increased conversion rates and revenue growth. By streamlining its checkout process and cutting the average time to completion by 20%, a retailer might see a 35% increase in transactions, translating to millions in additional revenue. This improvement signals positive trends in user behavior long before financial reports reflect the full impact.

These predictive insights are not limited to e-commerce. Let's examine other examples of UX success metrics that serve as early indicators of business performance and their relationship to lagging KPIs.

Improving Operational Efficiency with Internal Tools

Higher usability scores in internal tools often signal improved employee productivity and reduced operational costs. Imagine a financial services firm revamping its loan processing software to streamline workflows. By cutting task completion times by 40%, the organization could save hundreds of employee hours each week while enhancing overall job satisfaction.

Enhancing Retention Through Onboarding Improvements

Customer satisfaction during onboarding is a strong predictor of retention and long-term value. Consider a SaaS company redesigning its onboarding process to address user pain points. If satisfaction scores increased by 15% and churn dropped by 20%, the company would retain more customers and significantly boost revenue through higher lifetime value.

Reducing Drop-Off Rates to Drive Engagement

Monitoring drop-off rates in product usage can uncover friction points that hinder engagement. For example, if a music streaming platform identifies a high drop-off rate, it could simplify its playlist creation interface. This change could lead to an increase in feature usage, improving overall retention and deepening user engagement.

Adoption of Self-Service Portals

Self-service portals are a valuable UX metric with clear business implications. Imagine a telecommunications provider launching an intuitive self-service portal. By empowering users to resolve issues independently, the company could reduce the need for customer support calls by 40%, achieving significant cost savings while enhancing customer satisfaction.

Prototyping as a Predictor of Product Success

Prototype usability testing provides critical insights into a product's potential success. Imagine a tech company gathering user feedback on a prototype dashboard. The company could achieve a 30% higher adoption rate by addressing usability issues before launch, ensuring a smoother rollout and greater overall success.

Why Businesses Need Leading Indicators

UX success metrics provide something uniquely valuable to business leaders: proactive insight. Unlike traditional lagging indicators such as profit margins, customer retention rates, or employee turnover, which offer essential snapshots of past performance, UX metrics serve as leading indicators. These metrics deliver real-time data on user behaviors, satisfaction, and engagement, empowering leaders to act before issues escalate. By addressing challenges early, businesses can prevent costly problems, allocate resources more effectively, and align teams around shared goals.

For instance, early usability testing during the development of a customer onboarding process might reveal high error rates that could frustrate users and increase churn. Acting on this insight before launch allows teams to refine the design, ensuring a smoother rollout and better user retention. Similarly, tracking the adoption and satisfaction rates of a pilot self-service portal can guide leadership to confidently scale the initiative, knowing it will deliver a strong return on investment.

UX metrics also play a critical role in fostering cross-functional alignment. By providing actionable, measurable goals, such as improving task completion rates or reducing drop-off points, these metrics become a common language for collaboration across departments like product, marketing, and operations. This shared focus ensures every team contributes to delivering a seamless and effective user experience.

By providing actionable, measurable goals, these metrics become a common language for collaboration across departments.

Moreover, tracking usability scores during the development of internal tools can prevent wasted time and resources. For example, a company creating a dashboard for managing customer inquiries might uncover, through early testing, that employees struggle to find key features, leading to slower response times. Identifying and addressing these issues during the design phase avoids the operational inefficiencies and employee frustration that would have arisen from a poorly implemented tool.

In today's fast-paced business environment, maintaining strategic agility is critical. Leading indicators like UX metrics allow organizations to adapt quickly, optimize their investments, and deliver solutions that meet user needs from the outset. For CEOs and leadership teams, leveraging these insights is not just a tactical advantage but a strategic imperative. By acting on these early signals, businesses position themselves for sustainable growth, resilience, and long-term success.

Realizing the impact of investment in UX

Understanding the connection between UX success metrics and business KPIs unlocks a powerful strategic advantage. These metrics act as an early warning system, identifying opportunities to enhance user experiences, reduce friction, and drive growth long before traditional lagging indicators reflect the outcomes.

When leadership embraces UX metrics as leading performance indicators, the organization shifts from reactive problem-solving to proactive decision-making. By addressing challenges early and aligning UX efforts with measurable outcomes like revenue growth, cost savings, retention, and productivity, organizations lay the groundwork for sustainable growth and resilience. Next, we'll explore aligning UX with broader business strategy, ensuring that every initiative contributes meaningfully to organizational success.

NOTES

NOTES

Chapter 3:
Aligning UX with Business Strategy

Design should not be a bunch of designers who sit in the corner and design stuff, and then magically, good things happen. You have to have design as a central and collaborative piece that infects and impacts all aspects of the business.

John Maeda
VP Engineering, Head of Computational
Design Microsoft

Business leaders understand the critical role of alignment. Your company thrives when your teams, resources, and initiatives work together toward a common goal. Yet too often, UX is treated as a tactical afterthought, an isolated design detail or a tool for fixing immediate issues. This narrow view limits UX's potential to drive meaningful outcomes.

When integrated with business strategy, UX bridges the gap between high-level objectives and tangible results, designing systems and interactions that empower people to achieve their goals. Whether it's customers navigating your website or employees using internal tools, UX translates strategic intent into measurable outcomes. Companies that embrace this connection unlock a significant competitive advantage.

The Problem with Friction

Friction, whether manifesting as confusing layouts, inefficient workflows, or disconnected systems, can derail even the most well-crafted strategic plans. These barriers create unnecessary complexity, frustrate users, and ultimately dilute organizational efforts. UX addresses these challenges by identifying obstacles, simplifying complexity, and crafting seamless solutions. The result is a transformation of strategic intent into actionable outcomes.

Responding to Evolving Customer Expectations

In an era where customer expectations in the financial sector are rapidly evolving, Grupo Financiero Banorte, one of Mexico's largest financial institutions, recognized the growing demand for seamless and accessible banking experiences. To meet this demand, Banorte launched Bineo, Mexico's first fully digital bank, aimed at bridging the gap between traditional banking and the digital-first expectations of modern customers.[13]

Designing for Engagement

Banorte's goal with Bineo was ambitious: to create a banking platform that catered to tech-savvy users yet remained accessible to those with varying levels of financial literacy. As fintech startups continued to disrupt the market with innovative solutions, Banorte focused on crafting an intuitive, secure, and engaging digital environment. This involved simplifying complex banking tasks with intuitive interfaces and incorporating modern design principles to enhance the user journey, thereby making digital banking easier for all customers.

Transformative Results and Market Impact

The introduction of Bineo resulted in transformative outcomes for Banorte. Within just three months of its launch, Bineo achieved a 669% increase in its customer base, growing to 10,000 users. This rapid growth was supported by high retention rates and customer praise for the platform's simplicity and reliability. By moving traditional banking processes online, Banorte significantly enhanced operational efficiency and set a new standard for digital banking in the region.

The success of Bineo was largely due to Banorte's strategic emphasis on user experience. By prioritizing user needs and integrating extensive customer feedback throughout the development process, Banorte ensured that Bineo not only met but exceeded user expectations. This commitment to innovation and customer-centric design has established Banorte as a leader in the financial sector, transforming the banking experience in Mexico and highlighting the critical role of UX in achieving customer satisfaction and operational excellence.

Connecting UX to Business Goals

The quality of user experience directly shapes revenue growth, operational efficiency, customer loyalty, and employee satisfaction. Here's how UX directly supports critical business goals:

Driving Revenue Growth

UX plays a vital role in boosting conversion rates and reducing friction in purchasing processes. Simplifying these interactions can lead to significant financial impact. For example, Amazon's "1-Click" checkout revolutionized e-commerce by removing unnecessary steps, dramatically increasing revenue. Similarly, Netflix's recommendation engine leverages personalization to enhance engagement and increase lifetime customer value.

Enhancing Operational Efficiency

Streamlined workflows enabled by thoughtful UX save time and reduce errors, improving overall operational efficiency. A redesigned CRM system, for instance, can free up sales teams from tedious data entry, allowing them to focus on closing deals. FedEx optimized its package-tracking tools for delivery drivers, improving route efficiency and significantly reducing costs. These UX improvements eliminate bottlenecks, enabling teams to work smarter and faster.

Building Customer Retention and Loyalty

Intuitive, seamless experiences are essential for fostering repeat business and loyalty. Apple's integrated ecosystem exemplifies this principle, providing frictionless interactions across devices and ensuring customers stay engaged within its ecosystem. Likewise, improvements to customer support channels, such as self-service options and streamlined processes, enhance satisfaction while reducing churn. UX isn't just about satisfying users in the moment. It's about keeping them coming back.

Improving Employee Productivity and Satisfaction

Internal UX is just as critical as customer-facing experiences. Efficient tools and systems directly influence employee morale and productivity. IBM's overhaul of its outdated internal tools saved employees thousands of hours by streamlining workflows and eliminating redundancies. This boosted employee satisfaction and reinforced the connection between operational KPIs and user-centered improvements.

The Strategic Value of UX

When UX is aligned with KPIs like conversion rates, churn reduction, cost savings, and employee engagement, it becomes a measurable driver of sustainable growth. By addressing the needs of both external and internal users, UX strengthens the entire business system, delivering value across every touchpoint. This alignment is about creating a cohesive approach that ties UX efforts directly to strategic outcomes, ensuring long-term success for your organization.

Bridging the Trust Gap

Airbnb's journey from a fledgling startup to a global leader
in hospitality is a testament to the strategic power of UX. In
its early days, Airbnb faced a fundamental challenge: a lack
of trust between its two key user groups. Hosts were wary of
letting strangers into their homes, and guests were concerned
about the safety and quality of accommodations. This trust
gap posed a significant barrier to adoption, threatening the
company's ability to scale.

Airbnb's UX team recognized the need to address the trust
gap between hosts and guests, launching a series of targeted
initiatives designed to build confidence at every stage of the
user journey. They introduced verified profiles that required
government ID verification, adding a layer of credibility and
reassuring users of the legitimacy of those they interacted
with. Airbnb implemented a detailed review system to foster
transparency and accountability, allowing hosts and guests
to leave feedback. This empowered users to make informed
decisions and encouraged responsible behavior on both sides of
the transaction.[14]

*By addressing the needs of both
external and internal users, UX
strengthens the entire business system.*

The company also developed a secure payment system that
held funds until after the stay was completed, eliminating
financial risks and instilling trust. To address concerns

about misleading property descriptions, Airbnb offered free professional photography services to hosts, ensuring that listings accurately represented the accommodations. Finally, Airbnb introduced secure, in-platform messaging, allowing hosts and guests to communicate directly before confirming a booking. This feature enabled users to ask questions, clarify expectations, and build rapport, further strengthening trust. These initiatives addressed the trust gap and became integral to Airbnb's growth and success.

Unleashing Growth

These UX-driven changes had a profound effect. Hosts felt more confident opening their homes, and guests gained the assurance needed to book accommodations in unfamiliar locations. By eliminating key barriers to adoption, Airbnb unlocked massive growth potential, transforming from a niche service into a global platform operating in over 220 countries and regions.

The results were remarkable. Trust became a cornerstone of Airbnb's user experience, driving increased adoption rates and rapid expansion of its user base. These efforts not only solidified Airbnb's reputation but also helped establish it as a defining force in the sharing economy. Today, Airbnb is valued at over $100 billion, a testament to the power of UX to address critical business challenges and fuel sustained growth. By aligning user trust with strategic goals, Airbnb demonstrated how thoughtful design can turn potential barriers into opportunities, setting the stage for long-term success.

From Design Function to Core Business Capability

To fully realize its potential, UX must be understood as a core business capability that drives strategic goals, not merely as a design function focused on aesthetics or screens. Integrating UX at the strategy level ensures that the user perspective shapes the products, services, and tools that define your business, whether the user is a customer navigating your website or an employee managing a critical workflow.

UX influence should extend across nearly every aspect of your organization. It touches product development, sales, marketing, and operations, working as a unifying force to create a cohesive, high-performing system that benefits both customers and employees. Skilled UX practitioners are essential in this process. They design user-centered, actionable solutions tied to measurable outcomes, ensuring UX efforts directly contribute to business success.

Bridging UX and Business Goals

To ensure UX drives meaningful results, it must be intentionally aligned with your broader business strategy. This alignment transforms UX from a creative exercise into a deliberate contributor to your company's objectives. Here's how to integrate UX into your strategic framework:

Tie UX Metrics to Business KPIs

Defining success for UX requires linking user outcomes directly to business results. For example, if customer retention is a priority, measure how improvements to onboarding or support processes reduce churn. For operational efficiency, track how redesigned internal tools save time or reduce employee errors. Metrics such as Net Promoter Score (NPS), customer satisfaction (CSAT), or task completion rates can provide valuable insights into the effectiveness of UX initiatives. By aligning these metrics with broader business KPIs, you ensure that UX efforts contribute directly to the outcomes that matter most.

Involve UX Practitioners in Strategic Planning

It is critical to integrate UX practitioners into early discussions about product roadmaps, market positioning, and business priorities. Their expertise in user behavior and pain points provides invaluable insights that can guide key decisions. For instance, UX teams can help prioritize features that address customer needs, identify emerging trends, or recommend where to invest in tools and technologies. Including UX professionals in these conversations ensures that user perspectives are considered throughout the planning process, leading to decisions more aligned with user and business goals.

Use Journey Mapping to Identify Opportunities

Journey mapping is more than a tool for improving experiences, it's a method for identifying alignment or friction within your business. By visualizing the customer or employee journey, organizations can uncover where user frustrations overlap with inefficiencies, pinpoint critical touchpoints that drive loyalty or revenue, and identify gaps in service delivery that impact both users and the bottom line. Journey maps connect the dots between user needs and organizational goals, helping businesses prioritize the areas that will significantly impact performance and satisfaction.

Leverage UX Insights to Guide Innovation

UX research is a powerful tool for uncovering emerging trends and unmet user needs, providing a clear roadmap for innovation. These insights can validate new product concepts before significant resources are committed to development, ensuring efforts are focused on solving real problems. UX research can also identify areas where competitors fall short, allowing your offerings to fill those gaps and differentiate your business. By testing ideas with users early in the process, you can refine solutions and mitigate risk, ensuring that innovation is grounded in user research rather than assumptions.

Create a Feedback Loop Between UX and Leadership

Regular communication between UX teams and senior leadership is essential for aligning efforts with business priorities. Establish structured touchpoints where UX teams present their findings, highlight progress on initiatives, and share outcomes tied to business metrics. These sessions should also focus on emerging user trends that inform future strategy and provide actionable recommendations for investments or adjustments. This feedback loop keeps UX aligned with the organization's goals while empowering leadership with insights that drive decision-making.

Align UX with Cross-Functional Goals

Embedding UX into your strategy requires close collaboration across marketing, sales, product, and operations departments. For example, UX teams can work with marketing to ensure the website aligns with brand messaging and drives conversions. They can partner with product teams to prioritize features that deliver the highest value to users and collaborate with HR to improve employee tools and workflows, enhancing productivity and morale. By tying UX efforts to the goals of these teams, you create a cohesive system where user-centered design amplifies the effectiveness of every department.

Scale UX Practices Thoughtfully

As your UX practice grows, ensure it scales in alignment with your business objectives. This may involve hiring dedicated UX leadership to guide strategy and advocate for user-centered design at the executive level. Investing in tools and training is equally important to enable teams to deliver consistent, high-quality experiences. Standardizing key processes like usability testing or prototyping ensures efficiency and alignment across projects, allowing your UX efforts to evolve alongside the broader organization. Scaling thoughtfully ensures that UX continues to deliver value as your business grows.

Catalyst for Organizational Success

By aligning UX with your business strategy, you move beyond creating better products or services. You create a system where user-centered thinking drives creates impact. This approach strengthens the role of UX within your organization and positions your business to thrive in a competitive, user-driven marketplace.

Next, we'll explore the critical relationship between UX and CX. Understanding this connection is essential for delivering cohesive, impactful experiences that align with your business goals.

NOTES

NOTES

Chapter 4:
The UX and CX Connection

You've got to start with the customer experience and work backwards for the technology. You can't start with the technology and try to figure out where you're going to try to sell it.

Steve Jobs

UX and CX are often discussed in tandem, and with good reason, they are deeply interconnected. CX encompasses a customer's entire journey with your brand, from their first interaction to their long-term loyalty. UX, on the other hand, zeroes in on the usability and functionality of specific interactions within that broader journey. If CX is the story, UX is how the pages turn.

A successful CX strategy depends on seamless, well-designed UX touchpoints that smoothly move the customer from one stage to the next. Conversely, even the most polished UX can fall short if the overarching CX strategy is fragmented or poorly executed. Aligning these two disciplines is essential for delivering cohesive, satisfying customer journeys.

Investing in customer-facing UX is more than a competitive advantage; it's necessary in today's market. Leading companies recognize this and consistently prioritize the quality of the customer experience to build loyalty and trust through intentional, thoughtful design.

If CX is the story,
UX is how the pages turn.

According to Forrester's 2025 Budget Planning Guide for Customer Experience, 40% of global CX leaders plan to increase their CX investments above inflation within the following year.[15] This trend highlights the growing understanding that superior customer experiences directly correlate with business success.

McKinsey's research indicates organizations with strong alignment among their marketing, digital, and CX teams report 1.6 times faster revenue growth and 1.4 times better customer retention than their peers.[16] These figures underscore the importance of cohesive, cross-functional efforts in creating connected, impactful experiences.

When businesses invest in customer-facing UX, they enhance satisfaction and position themselves for sustainable growth and resilience in an increasingly competitive landscape. UX ensures that every interaction, from a customer's first click to their post-purchase engagement, feels intuitive and frictionless. This is what turns satisfied users into loyal advocates and transforms strong products into exceptional brands.

How UX Enhances CX

Every customer journey is made up of countless interactions, small moments that can either build trust or create friction. UX focuses on optimizing these moments, ensuring they are intuitive, frustration-free, and seamlessly aligned with the broader goals of Customer Experience (CX). Across industries, UX plays a pivotal role in transforming individual touchpoints into cohesive, satisfying journeys.

Retail

In retail, CX encompasses the customer's entire shopping experience, from seeing an ad to receiving their order. UX refines each touchpoint along the way, navigating the site, selecting a product, and completing checkout, ensuring the process is seamless and efficient. Features like auto-filled payment information or real-time inventory updates eliminate common barriers, reducing cart abandonment and driving conversions.

Travel

CX spans the entire travel journey, from booking and airport check-ins to flights and destination services. UX ensures critical touchpoints, such as intuitive booking platforms, mobile boarding passes, and real-time flight status updates, are smooth and stress-free. For instance, a well-designed rebooking interface during flight delays can maintain trust and satisfaction, while a poor experience can quickly erode both.

Healthcare

In healthcare, CX involves everything from scheduling appointments to accessing test results and follow-up care. UX optimizes these interactions through user-friendly portals that allow patients to easily find doctors, book visits, and access clear, actionable health data. Features like appointment reminders or chat tools for quick questions make the process proactive and reassuring, improving patient satisfaction and engagement.

Streaming Content

CX includes content enjoyment, pricing plans, and customer support for streaming platforms like Netflix or Spotify. UX enhances these experiences by providing personalized recommendations, effortless navigation, and seamless playback controls. A poorly designed interface or irrelevant recommendations disrupt the user experience, reducing satisfaction and retention.

Banking

CX encompasses the customer relationship in banking, including branch visits, mobile app interactions, and support services. UX ensures routine actions like checking balances, transferring funds, or disputing charges are simple and intuitive. Features like real-time spending alerts or personalized savings tips create a sense of security and value, fostering loyalty and trust.

Hospitality

CX includes researching destinations, booking stays, checking in, and accessing guest services in hospitality. UX enhances these interactions with smooth booking engines, efficient check-in processes, and user-friendly digital room controls or concierge services. Frustration-free touchpoints elevate satisfaction and encourage repeat stays.

Driving CX Success

Across industries, UX serves as the engine that powers great CX by removing friction, simplifying interactions, and leaving a lasting positive impression. By optimizing individual touchpoints, UX transforms the broader customer experience into one that builds loyalty, reduces churn, and enhances your brand's reputation.

Revolutionizing Air Travel

Delta Air Lines has become a leader in leveraging UX enhancements to elevate CX, with the Fly Delta app playing a crucial role in these efforts. Air travel is fraught with potential stressors like flight delays, last-minute gate changes, lost baggage, and long queues at check-in counters. Recognizing these challenges, Delta has transformed its digital capabilities into empowering tools for passengers, enhancing operational efficiency and fostering loyalty.[17]

Core Features and Benefits of the Fly Delta App

The Fly Delta app serves as a cornerstone of Delta's strategy to enhance the customer journey by mitigating common travel frustrations with innovative digital solutions. Key features of the app include:

Real-time Updates: Passengers receive instant notifications about gate changes, delays, and baggage status, reducing uncertainty and easing travel anxiety.

Mobile Boarding: The app enables seamless check-ins and access to digital boarding passes, eliminating the need for paper tickets and expediting the boarding process.

Efficient Rebooking: In the event of delays or cancellations, the app offers easy rebooking options, allowing passengers to avoid long lines at service counters.

Baggage Tracking: Integrated with Delta's RFID system, the app provides real-time luggage location updates, giving travelers peace of mind.

Loyalty Integration: SkyMiles members can track rewards, view exclusive offers, and manage their accounts, enhancing user engagement and loyalty.

Operational Efficiency and Customer Satisfaction

Delta's investment in the app has reaped significant benefits, both for travelers and the airline itself. In 2023, the app recorded over one billion unique engagements, illustrating its critical role in the customer journey. Features like real-time updates and intuitive navigation have markedly improved customer satisfaction, with passengers feeling more informed and in control. Operational efficiency has also been boosted, as the self-service features reduce the workload on airport staff and customer support teams, leading to fewer support calls and shorter lines at airports.

Strategic Impact and Industry Recognition

The app's impact extends to enhancing customer loyalty by increasing engagement with the SkyMiles program, deepening customer relationships, and driving repeat bookings. Delta's focus on reliability and efficiency gained further recognition when the airline was named the most on-time North American carrier in 2022. This accolade highlights the seamless integration of digital tools like the Fly Delta app into Delta's operational strategy.

Setting a New Standard in Air Travel

The Fly Delta app is more than just a convenience. It is an integral part of Delta's strategy to stay competitive in an increasingly digital and customer-focused industry.[18] By prioritizing UX, Delta has not only enhanced the overall travel experience but also reinforced its reputation for customer-centric innovation, turning common travel frustrations into opportunities for strengthening loyalty, efficiency, and brand reputation.

How UX Practitioners Transform CX

UX practitioners play a pivotal role in shaping and improving CX. Their work goes far beyond aesthetics, focusing instead on addressing the real pain points and opportunities within the customer journey. By grounding their efforts in research and collaborating across teams, UX practitioners ensure that every interaction contributes meaningfully to a cohesive, satisfying experience. These are not just tactical fixes but strategic interventions designed to align with broader CX and business goals.

At their core, UX practitioners aim to create experiences that remove friction, foster trust, and deliver value. This is achieved through deliberate activities that bridge the gap between user needs and organizational priorities. Using tools like journey mapping, iterative prototyping, and usability testing, UX practitioners deliver actionable insights and practical solutions that enhance CX at every journey stage. This collaborative, research-driven approach positions UX practitioners as essential partners in creating exceptional customer experiences. By precisely addressing specific touchpoints and aligning their activities with strategic goals, they enhance CX and contribute directly to business performance. Let's look at some of the specific methods and tools they use to make this happen:

Usability Testing: Observing Real Users

Usability testing involves observing how real users interact with a product or system to identify pain points and areas for improvement. This process highlights challenges that may not be obvious from data alone. For instance, users might abandon a form due to unclear instructions or overly complex steps. By catching these issues early, teams can make adjustments that improve usability. boost engagement, and ensure a more seamless experience.

Prototyping and Iteration: Testing Before Scaling

Prototyping allows UX teams to test and refine ideas before committing to full implementation. Prototypes provide a low-risk way to validate solutions and gather feedback, whether it's a feature update or a process redesign. Iterative design ensures that the final product meets user expectations while avoiding costly missteps. For example, a prototype for a new navigation menu might reveal usability gaps that can be addressed early, resulting in a solution that's intuitive, effective, and ready to scale.

Data Analysis: Uncovering Behavior Patterns

By utilizing tools like heatmaps, user session recordings, and behavioral analytics, UX professionals identify user behavior patterns and pinpoint areas where users disengage. This data-driven approach ensures that design decisions are based on actual user interactions rather than assumptions. For instance, analyzing user data might reveal that customers abandon a website during checkout due to high shipping costs or a confusing interface. With these insights, UX teams can implement targeted changes to address these issues, thereby improving conversion rates and customer satisfaction.

Cross-Functional Workshops: Aligning Teams Around the User

Facilitating workshops that bring together departments such as marketing, operations, product development, and customer support, UX practitioners promote collaboration focused on user needs. These sessions ensure that all teams contribute to a cohesive customer experience. Research indicates that organizations excelling in cross-functional collaboration are more likely to achieve significant revenue growth compared to those operating in silos. By fostering open communication and aligning departmental goals, UX professionals help dismantle internal barriers, leading to a unified CX strategy.

Design Thinking: A Framework for Innovation

Many UX practitioners adopt design thinking, a user-centered, iterative approach emphasizing empathy, ideation, and prototyping, to tackle complex CX challenges. This methodology encourages innovative solutions tailored to user needs. For example, companies that integrate design thinking into their processes have reported substantial returns on investment, with some projects achieving up to 229% ROI. By focusing on understanding the user experience, organizations can develop products and services that resonate more deeply with their audience, enhancing overall customer satisfaction.

Why These Activities Matter

The activities carried out by UX practitioners are tactical and transformative. They bridge the gap between specific user interactions and the broader customer experience, ensuring that every touchpoint contributes meaningfully to the success of the brand. By identifying and resolving pain points, UX practitioners improve individual interactions and elevate the entire customer journey. Strategic practices like journey mapping, usability testing, and data analysis don't just benefit users. They drive value for the entire organization. They help align CX efforts with measurable business outcomes, ensuring that every improvement contributes to metrics that matter.

Delivering a cohesive customer experience requires breaking down silos and fostering collaboration across teams. Too often, CX strategies are driven by sales, marketing, or customer service, while UX efforts remain confined to product or design teams. This separation creates disjointed experiences where customer pain points are overlooked, or solutions fail to achieve their full potential. When UX and CX teams align around a shared focus on the customer, they can design seamless, intentional, and satisfying journeys. This collaboration ensures that insights from every department contribute to a unified experience that addresses pain points holistically.

The key to achieving this alignment lies in building bridges between these functions, creating opportunities for cross-functional collaboration that keep the user at the center. Lets explore practical strategies for bringing teams together to drive alignment and deliver exceptional customer experiences:

Establish Shared Goals

Bridging the gap between UX and CX starts with aligning teams around shared objectives. Define CX metrics that UX efforts can directly influence, such as Net Promoter Score (NPS), customer retention rates, or time-to-resolution for support inquiries. For example, if the goal is to improve customer retention, UX might focus on refining the onboarding process to make it more intuitive or simplifying self-service options to empower users. These shared goals provide clarity, ensuring every team understands how their work supports the broader customer journey.

Foster Communication Across Teams

Breaking down silos requires open, consistent communication across departments. Regular cross-functional meetings bring together UX, CX, marketing, product, and customer service teams to share insights, challenges, and priorities. These sessions encourage collaboration and uncover opportunities for alignment, like how a marketing campaign impacts the design of a landing page or how UX improvements in support tools can enhance customer interactions. By fostering these conversations, teams can work together to deliver seamless, connected experiences that benefit both users and the business.

Create Feedback Loops for Continuous Improvement

Customer feedback is a powerful resource for refining UX touchpoints and broader CX strategies. By creating feedback loops that leverage tools like surveys, usability tests, and customer support data, teams can identify and address real problems collaboratively. For instance, if CX teams report frequent customer complaints about a confusing feature, UX can redesign it to reduce frustrations and improve usability.

Conversely, usability testing might reveal broader issues in the customer journey, like difficulty finding critical information, that CX teams hadn't considered. These feedback loops create a continuous cycle of learning and improvement, ensuring teams stay responsive to evolving customer needs.

Build a Shared Understanding of the Customer Journey

Journey mapping exercises are a practical way to bring teams together and visualize the customer experience from initial awareness to post-purchase engagement. By including input from stakeholders across departments like sales, support, and UX design, journey maps can reveal gaps or redundancies and highlight opportunities for alignment. For example, a map might show that while a website's checkout process is seamless, post-purchase communications fail to set clear delivery expectations. Armed with this insight, UX and CX teams can collaborate to design solutions that close the gap, ensuring a smoother and more satisfying experience for the customer.

Leverage Cross-Functional Metrics

Organizations that align UX and CX efforts gain a strategic advantage by focusing on cross-functional metrics that reflect both user satisfaction and business outcomes. Metrics like Net Promoter Score (NPS), customer retention rates, and conversion rates provide a shared framework for success. Companies that excel in this alignment report faster revenue growth and better customer retention, demonstrating the impact of collaboration across marketing, digital, and CX teams. These metrics help teams see the bigger picture and clarify how their individual contributions support the organization's goals.

The Power of UX and CX Alignment

By establishing shared goals, fostering communication, and creating a culture of collaboration, businesses can align UX and CX teams to design seamless, intuitive customer journeys. This alignment doesn't just enhance the user experience; it delivers measurable results, from higher retention rates to increased revenue. When teams work together toward a shared vision, the entire organization benefits, customers receive exceptional experiences, employees are empowered, and the business thrives.

When UX and CX work in harmony, the results are transformational.

The business, in turn, benefits from higher customer satisfaction, increased retention, and measurable growth. UX refines the touchpoints; CX defines the strategy.

Together, they create the kind of cohesive, impactful experiences that customers remember, and competitors struggle to replicate. However, the impact of UX doesn't stop with external users. In the next chapter, we'll explore how designing for employees, through better tools, workflows, and systems, unlocks internal efficiencies and creates a ripple effect that strengthens the customer experience.

NOTES

NOTES

Chapter 5:
The Power of Internal UX

Clients do not come first. Employees come first. If you take care of your employees, they will take care of the clients.

Richard Branson
CEO and founder—Virgin Group

When business leaders think about UX, they often focus on customer-facing experiences. It's a logical starting point, after all, customers drive revenue. However, this narrow view overlooks a significant opportunity to improve productivity, morale, and satisfaction: the employee experience. Employees' interaction with internal systems, tools, or workflows is part of their user experience. Whether they're submitting an expense report, navigating a CRM, or accessing an intranet, these interactions shape how employees feel about their work and how efficiently they can perform it. Yet, internal UX often falls by the wayside, leaving employees to contend with clunky interfaces, outdated software, and cumbersome processes.

Neglecting internal UX has consequences far beyond frustration. Employees burdened by poorly designed tools take longer to complete tasks, feel less engaged in their roles, and are more likely to leave. These inefficiencies also ripple outward, impacting customer interactions. How many minutes have you personally spent on hold while customer service deals with a slow computer? Slower response times, errors, and diminished service quality often stem from employees' challenges behind the scenes.

The Hidden Cost of Bad Internal UX

Beyond customer inconvenience, the impact of poor internal UX can be devastating. Consider the case of Knight Capital Group, a financial services firm that, on August 1, 2012, lost $460 million in just 45 minutes due to a software error.[19] The root cause was a misconfigured software deployment in their trading system, where employees relied on overly complex and poorly integrated tools, creating an environment where critical mistakes could, and did, happen.

While Knight Capital's collapse is an extreme example, it highlights the risks of neglecting internal UX. Most organizations won't face a catastrophe of this scale. Still, the hidden costs of bad internal UX are pervasive. Employees struggling with clunky interfaces, outdated workflows, or disjointed systems take longer to complete tasks, make more errors, and feel increasingly disengaged. These inefficiencies silently erode productivity, drive up operational costs, and contribute to employee turnover.

Neglecting internal UX has consequences far beyond frustration.

Worse, the ripple effects often extend to customers, resulting in slower response times, lower service quality, and missed opportunities to build loyalty. The cost isn't just financial, it's cultural. Frustrated employees are less likely to feel invested in their work, and the toll on morale can undermine collaboration and innovation across teams. Let's look at how businesses that prioritize internal UX are turning these challenges into opportunities. These examples show how thoughtful design can transform the employee experience, driving efficiency, engagement, and better outcomes for the entire organization.

Inefficient Systems Leading to Productivity Loss

Target has been actively investing in modernizing its inventory management systems to enhance efficiency.[20] For instance, the company has implemented a new inventory control system to better position inventory within its network, streamline backroom operations, and maintain optimal in-stock levels. These improvements have enabled store teams to focus more on guest-facing work, thereby increasing overall productivity.

Impact on Employee Turnover

Kaiser Permanente recognized the challenges posed by outdated scheduling systems and has been transitioning to new workforce management technology after using the same scheduling system for three decades.[21] This shift is part of their efforts to improve employee satisfaction and operational efficiency.

Errors and Compliance Risks in Healthcare

Cedars-Sinai Medical Center conducted a study indicating that when pharmacy professionals, rather than doctors or nurses, take medication histories of patients in emergency departments, mistakes in drug orders can be reduced by more than 80%.[22] This finding underscores the importance of accurate medication histories in preventing errors and enhancing patient safety.

Customer Frustration Stemming from Poor Internal Tools

Delta Air Lines has been investing in upgrading its customer service tools to enhance efficiency and improve passenger satisfaction. For instance, the airline has focused on change management in CX technology rollouts to ensure that their staff effectively integrates and utilizes new systems.[23] These efforts aim to reduce customer service handling times and provide a more seamless experience for passengers.

Missed Opportunities for Innovation and Growth

Intel has recognized the importance of efficient project management tools in fostering innovation. The company has implemented systems to improve process management and collaboration, enabling teams to work more effectively and focus on strategic initiatives.[24] By streamlining workflows and reducing time spent on administrative tasks, Intel has been able to allocate more resources toward product development and innovation.

Empowering Employees Through Internal UX

Southwest Airlines has built its reputation as a customer service leader by prioritizing its employees' needs. Recognizing that empowered employees drive exceptional customer experiences, the company has invested heavily in modernizing its internal systems to create a seamless and efficient work environment. They faced significant operational bottlenecks due to fragmented tools and outdated systems. Employees, especially frontline gate agents and flight crews, struggled to navigate complex workflows and disjointed platforms. These inefficiencies often led to slower responses to customer needs, delayed rebookings during disruptions, and additional frustration for both employees and passengers.

They recognized the need to equip their employees with tools that streamlined workflows and empowered them to deliver timely, high-quality service. In partnership with Salesforce, the airline reimagined its internal systems to prioritize the employee experience.[25] The resulting platform integrated real-time data, simplified workflows, and provided actionable insights to enhance operational efficiency.

Streamlining Operations

A key feature of the redesign was an integrated dashboard that unified previously siloed tools, giving employees seamless access to critical information in a single interface. Mobile accessibility was another cornerstone of the system, enabling employees to manage tasks on the go, whether at the gate or onboard a flight. AI-powered tools added a layer of intelligence, offering predictive analytics that allowed employees to anticipate potential disruptions and proactively address issues. Automated processes further reduced manual, repetitive tasks, freeing up staff to focus on meaningful customer interactions. Together, these innovations created a comprehensive platform that not only enhanced the employee experience but also strengthened Southwest Airlines' ability to deliver exceptional service.

Southwest's investment in employee-centric UX delivered measurable improvements across several key areas. Employees reported a significant reduction in the time spent navigating systems, which allowed them to focus more on assisting customers efficiently.

Minimized delays, reducing the costs associated with reactive problem-solving.

With intuitive tools that minimized frustration, employee morale and engagement levels rose noticeably. Passengers also benefited, as faster rebookings, real-time updates, and proactive communication enhanced the travel experience and boosted overall satisfaction scores. Additionally, predictive tools and integrated workflows minimized delays, reducing the costs associated with reactive problem-solving and further strengthening operational performance.

By aligning its technology with employee needs, Southwest Airlines demonstrated the powerful connection between internal UX and external customer experience. Empowered employees, equipped with tools designed for their success, became better equipped to deliver the friendly and reliable service Southwest is known for.

Strategic Improvements Without Starting Over

Fortunately, Improving internal UX isn't typically about scrapping everything and starting over. Most organizations operate with a mix of custom-built software and third-party solutions, creating inherent limitations around customization and integration. These constraints demand a thoughtful approach that relies on creative problem-solving, tight collaboration with engineering, and a focus on addressing the most impactful pain points. A few tips on how to tackle internal UX improvements effectively:

1. Listen to Employees

Employees are your most valuable source of insights when it comes to identifying what's working, and what isn't. They are on the front lines, navigating tools and processes daily, and their feedback is essential for uncovering friction points. Organizations should conduct surveys, interviews, and workshops to understand their pain points, asking targeted questions like: Which tools cause the most frustration? What tasks take longer than they should? Where do you feel unsupported in your daily work? These discussions not only highlight areas of inefficiency but also help prioritize improvements. By focusing on the challenges that employees face most frequently, you ensure that solutions address real, impactful needs and foster a culture of collaboration and trust.

2. Observe and Analyze Workflows

Understanding how employees interact with tools and processes requires more than just feedback, it demands observation and data. By using analytics tools and direct observation, organizations can uncover inefficiencies in workflows that might not be immediately apparent. For example, tracking the time it takes employees to complete routine tasks or navigate between systems can reveal bottlenecks. Patterns such as repeated steps, redundant data entry, or frequent errors often indicate unnecessary complexity that slows productivity and creates frustration.

In cases where systems offer limited customization options, collaboration with engineering teams becomes essential. Together, UX and engineering can explore alternatives such as integrating tools, automating repetitive actions, or reconfiguring workflows to eliminate inefficiencies. These approaches allow teams to make meaningful improvements without the need for costly, large-scale system overhauls. Observation and analysis provide the clarity needed to make targeted, effective changes that streamline operations and empower employees.

3. Prioritize Quick Wins

When improving internal UX, starting with small, high-impact changes can deliver immediate value while building momentum for more significant transformations. These quick wins often address the most visible and frustrating pain points, creating tangible improvements in employee satisfaction and productivity. Simplifying a frequently used form to reduce completion time, consolidating data from multiple systems into a unified dashboard, or automating repetitive tasks like data entry or report generation are all examples of changes that can yield outsized benefits.

By focusing on quick wins, organizations provide immediate relief to employees and demonstrate the potential of UX improvements. These early successes build trust and enthusiasm, laying the groundwork for tackling larger, more complex projects down the line. Quick wins show that even modest investments in UX can lead to meaningful gains, reinforcing the value of a user-centered approach across the organization.

4. Test and Iterate with Small Groups

Before implementing changes on a larger scale, piloting new tools or processes with a small group of employees is essential. This focused approach allows teams to gather real-world feedback, identify potential issues, and make adjustments before a full rollout. Collaborating closely with engineering during this phase ensures that technical limitations are addressed and the solution integrates seamlessly with broader system requirements.

Iterative testing builds confidence among employees and stakeholders by demonstrating that their input is valued and incorporated into the design process. By the time the solution scales, it's technically sound and optimized to meet employee expectations, ensuring a smoother transition and higher adoption rates across the organization.

5. Measure the Impact

To understand the effectiveness of your UX improvements, tracking key metrics that evaluate their impact is critical. Focus on indicators that reflect both employee productivity and satisfaction. For example, measure task completion times to see how long employees take to finish critical workflows before and after changes. Use employee satisfaction surveys to gauge how team members feel about the updated tools and processes. Additionally, turnover rates should be monitored to determine whether improvements in internal UX contribute to higher retention.

Maximizing Impact Within Constraints

Working with a mix of custom-built and third-party software often means you won't have full control over every system. However, meaningful improvements are still within reach. By collaborating closely with engineering, leveraging creative problem-solving, and focusing on the employee experience, you can design tools and workflows that reduce friction and unlock productivity, even when faced with existing constraints. Data-driven outcomes validate the success of UX initiatives, providing a compelling case for their value.

Seizing the Opportunity

The equation is simple: empowered employees deliver better outcomes. When tools and workflows are intuitive and efficient, employees can focus their energy where it matters most: delighting customers, driving innovation, and solving meaningful problems. Internal UX isn't a secondary concern or a luxury; it's a strategic investment that transforms operations, enhances employee satisfaction, and amplifies the quality of customer experiences.

Personal Experience

I've witnessed this transformation firsthand. At a Fortune 100 financial company, employees were burdened with more than a dozen disjointed systems, each critical but functioning independently. Routine tasks required navigating multiple platforms and manually entering data into custom spreadsheets, wasting time and increasing frustration. A full replatforming wasn't feasible, so we started with a more immediate solution: creating a unified dashboard. This dashboard brought disparate systems together into a single, cohesive interface. Employees could seamlessly access the tools and information they needed, saving time and reducing frustration. The results were tangible: greater efficiency, reduced errors, and a workforce empowered to perform at its best. For the business, the streamlined workflows unlocked significant value, reinforcing an essential lesson: even incremental improvements in internal UX can drive outsized results.

When businesses prioritize internal UX, they're not just enhancing tools. They're fostering an environment where employees thrive, operations excel, and customers reap the benefits of a more agile and capable organization. This focus on internal systems creates ripple effects that extend throughout the company, driving efficiency, satisfaction, and innovation. However, creating impactful UX requires more than just tools and processes. It demands a culture that champions user-centered thinking. In the next chapter, we'll explore how to build a UX-driven culture, embedding these principles into your organization to ensure sustained growth and a competitive edge.

NOTES

Chapter 6:
Building a UX Culture

The magic in any product or service is how it's experienced by the end user.

Phil Gilbert
former General Manager of Design—IBM

For many organizations, UX is already part of the business in some capacity. It might be embedded in product design, marketing, or customer support. However, whether you're just starting to introduce UX principles or looking to maximize the impact of an existing UX function, the goal is to ensure UX drives value across your organization. Integrating UX isn't about overhauling your entire business but refining how it operates. UX thrives when it complements and collaborates with other functions, aligning tools, services, and systems with the needs of the people who rely on them, whether they're customers or employees. By embedding UX into decision-making, you unlock its potential as a core capability that strengthens business outcomes.

As we've seen in numerous examples, a UX mindset equips your teams with insights and design solutions that align with strategic goals. It's not about elevating UX above other functions; it's about empowering all parts of the business to deliver better results by keeping the user perspective considered in every decision. When UX becomes an integral part of your strategy, it transforms outcomes across the board: from improving customer satisfaction to increasing operational efficiency and driving revenue growth.

Guiding Change

Embedding UX principles into an organization begins with leadership. Executives set the tone, and their commitment to user-centered thinking creates an environment where teams naturally integrate UX into their workflows. This doesn't require sweeping changes or costly initiatives. Often, small, consistent actions from leadership can drive meaningful transformation.

Leadership plays a pivotal role in fostering a UX-driven culture by providing direction, creating alignment, and modeling the importance of user outcomes. Here's how leaders can guide this change effectively:

1. Setting User-Centered Goals

Leaders are critical in connecting UX efforts to broader business objectives by establishing clear, measurable outcomes. By framing UX initiatives in terms of their impact on revenue, retention, or employee productivity, leadership ensures teams understand the strategic importance of user-centered thinking. For example, leadership might set a goal to reduce onboarding time for new customers by 20%, positioning it as both a user success metric and a business priority. This approach provides teams with a clear, actionable target and highlights the dual benefits for customers and the organization.

2. Encouraging Cross-Functional Collaboration

UX reaches its full potential when teams across product, marketing, sales, and operations work together to align their efforts around user needs. Leaders can encourage collaboration by breaking down silos and creating structured opportunities for alignment, such as cross-functional workshops, shared metrics, or regular discussions centered on user insights. For instance, a CMO might partner with the product team to ensure that marketing campaigns align seamlessly with website UX improvements, delivering a cohesive and effective user experience from the first ad click to final conversion. This kind of collaboration ensures that every touchpoint reinforces the organization's broader strategic goals.

3. Modeling the Approach

Leaders play a critical role in shaping a user-centered culture by demonstrating what it looks like in practice. Simple actions like asking thoughtful questions in meetings can refocus discussions on UX priorities. Questions like, "How does this feature help our customers achieve their goals?" or "What feedback have we received from users?" signal the importance of considering the end user in every decision. For example, during a product review, a VP might prioritize insights from usability testing over aesthetic preferences, reinforcing to the team that functionality and user satisfaction take precedence. By modeling this approach, leaders set the tone for user-centered decision-making across the organization.

4. Providing Resources and Support

Embedding UX into workflows requires a financial commitment to providing the necessary resources, tools, and talent. Leaders who prioritize UX ensure that teams have access to usability testing platforms, design tools, and training programs to strengthen their user-centered skills. For example, a business leader might sponsor a company-wide initiative introducing journey mapping workshops. These sessions equip teams with the tools and methodologies to collaboratively identify and address user pain points, fostering a culture of problem-solving and continuous improvement. By investing in these resources, leaders empower their teams to embed UX into everyday workflows effectively.

5. Creating Accountability Through Metrics

Leadership plays a crucial role in reinforcing the importance of UX by holding teams accountable for achieving user-centered goals. This involves tracking and reporting UX success metrics aligning with broader business KPIs. Regular reviews of these metrics ensure that UX remains a strategic priority. For instance, a CFO might evaluate task completion rates or user satisfaction scores alongside financial performance metrics during quarterly reviews. By drawing a clear connection between UX improvements and business outcomes, leadership emphasizes the tangible value of user-centered efforts and their impact on the organization's success.

By demonstrating that user needs are a key consideration in achieving business objectives, leaders help teams view UX as an essential part of the decision-making process. When leadership prioritizes UX, it sends a clear message: delivering exceptional experiences isn't just a function of design or product. It's a strategic imperative that drives the entire organization's success.

Elevating Digital Banking Through UX

Capital One has emerged as a leader in digital banking innovation by embedding UX principles into its organizational culture. Through a commitment to user-centered design, the company has transformed its mobile app into one of the most highly regarded platforms in the financial services industry. Capital One took a significant step in prioritizing UX in 2014 when it acquired Adaptive Path, a design firm known for its user-centered and service design expertise.[26] This acquisition allowed Capital One to integrate advanced design methodologies into its operations, fostering collaboration between product managers, engineers, and designers. The emphasis on aligning teams around the needs of users created a foundation for innovation and customer satisfaction.

Capital One's mobile app is a testament to the company's commitment to user-friendly functionality and intuitive design. The app includes a voice-activated assistant, Eno, which allows customers to manage accounts using natural language commands, enhancing convenience and accessibility. It also provides personalized insights, offering tailored spending recommendations and alerts that empower users with actionable financial information. Streamlined navigation further simplifies tasks like mobile check deposits, real-time transaction tracking, and bill payment, making everyday banking seamless.

*Aligning teams around the needs
of users created a foundation for
innovation.*

Capital One's investment in UX has yielded measurable results that underscore its impact on the digital banking landscape. According to app store data, by 2023, the Capital One Mobile app became one of the most downloaded banking apps in the U.S., achieving over 10 million downloads.[27] It maintains high user ratings, averaging 4.9 out of 5 stars on the Apple App Store and 4.5 out of 5 stars on Google Play, a testament to its consistent customer satisfaction.[28] As of December 2024, the app ranked #54 overall and #6 in the Finance category on the Apple App Store, highlighting its strong performance and competitive edge. Capital One's proactive approach to UX has strengthened customer satisfaction and engagement, solidifying its reputation as an innovator in digital banking. The company has set a high standard for the financial industry by integrating UX principles into its culture and operations.

Cross-Functional Collaboration

Great user experiences require coordination across multiple teams, including product development, marketing, sales, operations, and customer support. Effective cross-functional collaboration ensures that everyone understands how their work contributes to delivering seamless, impactful experiences. By aligning efforts, businesses create systems where UX insights inform decision-making across the organization.

Collaboration across functions is essential to embedding UX into an organization's strategy. Product and engineering teams benefit from partnering with UX to ensure that features are technically feasible, intuitive, and easy to use. Marketing and sales teams leverage UX insights to inform web design, messaging, and engagement strategies, ensuring campaigns effectively resonate with their target audience. Meanwhile, operations teams gain significant advantages from internal UX improvements, such as redesigned tools or streamlined workflows, which empower employees to deliver better results and enhance overall efficiency. This cross-functional alignment ensures that UX contributes value across the entire organization.

Regular collaboration through shared goals, workshops, and open communication channels ensures alignment and helps UX support every part of the organization. Marty Cagan's Product model provides a blueprint for creating an environment where cross-functional teams thrive, and UX delivers maximum value.

UX in the Product Operating Model

Marty Cagan, a renowned product leader and founder of the Silicon Valley Product Group (SVPG),[29] introduced his Product operating model to address the complexities of building successful products in modern organizations. Developed in the early 2000s and refined over decades, this model emphasizes integrating product management, UX, and engineering into a cohesive team. Many high-performing companies, such as Google, Amazon, and Spotify, have adopted versions of this model to create innovative, user-centric products while

achieving measurable business outcomes. The strength of
Cagan's model lies in its focus on collaboration, outcomes,
and empowerment. By aligning UX, product management, and
engineering, the model fosters an environment where teams
work toward shared goals, balancing user needs with technical
feasibility and business objectives.

UX and Product Management

In Marty Cagan's model, the partnership between UX and
product management (PM) is a cornerstone of effective product
development. Product managers focus on the "why" and
the "what" of the product, defining business objectives and
customer problems. At the same time, UX practitioners take
on the "how," ensuring solutions are intuitive, engaging, and
aligned with user needs. This collaboration drives an outcome-
focused approach where PMs set high-level goals, such as
increasing retention or reducing churn, and UX translates
these objectives into actionable designs. For instance, if churn
is linked to a confusing onboarding process, UX can create a
streamlined flow that guides users effectively and minimizes
frustration.

Additionally, UX practitioners validate these approaches
through user research methods like usability testing, journey
mapping, and analytics. This partnership reduces the risk
of costly missteps by grounding decisions in real-world
data rather than assumptions. UX and PM ensure product
development is tightly aligned with user needs and business
goals, fostering a shared commitment to delivering validated
product solutions.

UX and Engineering

The partnership between UX and engineering is critical to delivering products that meet user needs and technical feasibility. Engineers hold the keys to execution, ensuring that designs are implemented effectively while maintaining technical integrity. In Marty Cagan's model, this collaboration fosters alignment between design intent and practical implementation, bridging the gap between vision and reality. Engineers and UX practitioners share responsibility for quality, working together to ensure that the product delivers on its promise. For example, engineers might refine navigation flows or interaction details based on UX feedback, enhancing usability and preserving design fidelity during development. This shared commitment ensures the end product aligns with user expectations and technical standards.

A healthy UX practice also counterbalances engineering teams, which are often much larger and incentivized to prioritize speed and output volume. UX helps keep the focus on measurable outcomes, rather than simply shipping features quickly. This balance ensures the product delivers meaningful value, not just functionality. Creative problem-solving is another cornerstone of this partnership. Engineers bring technical expertise to address platform constraints, while UX practitioners contribute user-centered perspectives to simplify complex workflows or improve system efficiency. Together, they can develop innovative solutions that enhance usability without sacrificing technical feasibility.

This collaboration ensures the final product achieves both user satisfaction and technical excellence, driving measurable results and reinforcing the product's value to the business.

Why the Product Model Works

Marty Cagan's Product Org model creates an environment
where UX thrives by fostering collaboration, aligning goals,
and emphasizing outcomes that matter. By embedding UX
alongside product management and engineering, the model
ensures user needs remain a central focus throughout the
development process. For product management, UX delivers
the research and designs necessary to solve customer problems
effectively, translating high-level business goals into actionable
solutions. For engineering, UX provides a clear understanding
of user expectations, ensuring technical efforts are directed
toward creating meaningful, high-quality experiences.

This structure enhances cross-functional alignment, breaking
down silos and enabling teams to work together seamlessly. It
empowers UX to take on a strategic role, driving innovation,
improving customer satisfaction, and achieving measurable
business results. By integrating UX into the core of the product
organization, businesses can create solutions that meet user
needs while delivering real value to the organization.

Digital Transformation: A Unique Opportunity for UX

Digital transformation is reshaping businesses across
departments, offering organizations a rare chance to reimagine
their operations and the value they deliver. As companies adopt
new technologies, streamline workflows, and digitize customer
experiences, they create fertile ground for UX to demonstrate
its transformative potential. This shift to digital requires
collaboration across product teams, IT, marketing, operations,

and customer service, making UX a critical component
in ensuring that these efforts are not only technologically
advanced but also intuitive and impactful.

Simplifying Complexity

One of the primary challenges of digital transformation is
simplifying complexity. Organizations often implement
intricate systems like customer relationship management
platforms, AI-driven analytics tools, or enterprise resource
planning software. While powerful, these tools can overwhelm
users if they are not thoughtfully designed. Poorly designed
systems can increase friction, reduce productivity, and slow
adoption rates. By focusing on user needs, UX addresses these
challenges through intuitive interfaces, streamlined workflows,
and tools tailored to specific tasks. For instance, when a
logistics company introduced a new supply chain management
platform, users initially faced delays and errors due to a lack of
clear navigation. A UX-driven redesign simplified the interface,
integrated real-time updates, and automated repetitive tasks,
enabling users to concentrate on higher-value activities while
improving overall efficiency.

> *Ensuring that these efforts are not only*
> *technologically advanced but also*
> *intuitive and impactful.*

Resistance to new tools and technologies is another common
challenge. Many users are hesitant to abandon familiar
processes, which can hinder the adoption of digital systems.

UX mitigates this resistance by making systems user-friendly and providing onboarding experiences that build confidence and engagement. For example, a financial services firm launching a digital loan application platform ensured adoption by incorporating features like interactive tutorials, pre-filled data fields, and contextual help to guide customers through each step. These enhancements made the platform accessible and intuitive, reducing drop-offs and increasing customer satisfaction.

As organizations digitize, customers increasingly expect consistency across web, mobile, and in-person interactions. UX ensures that these touchpoints are seamlessly connected, fostering trust and strengthening loyalty. For example, a retail business integrating an online ordering system with in-store pickup must align the digital and physical experiences. A cohesive UX design might include real-time order tracking, clear pickup instructions, and the ability to apply rewards seamlessly, creating a frictionless experience that leaves a lasting positive impression.

The Impact of UX in Digital Transformation

Without a strong focus on UX, even the most advanced tools risk underutilization, leaving workflows inefficient and users disengaged. By addressing user needs, reducing friction, and creating seamless interactions, UX ensures that technology serves its intended purpose for customers. Prioritizing UX in digital transformation efforts delivers exceptional customer experiences and maximizes the return on technology investments. In an increasingly digital world, UX becomes the foundation that enables organizations to adapt, innovate, and thrive in competitive markets.

The Multiplier Effect of UX

The value of UX is magnified during digital transformation, creating ripple effects that enhance success across the organization. Embedding user-centered design into transformational projects accelerates adoption, strengthens customer relationships, and amplifies the overall impact of investments. Tools and platforms designed with UX principles produce faster and more effective results. For instance, a thoughtfully designed enterprise resource planning system can significantly shorten the learning curve, enabling users to work efficiently and realize value early.

Prioritizing usability and personalization

Digitized touchpoints that prioritize usability and personalization deepen customer loyalty. Seamless interactions, tailored experiences, and accessible designs foster trust and satisfaction, turning one-time customers into repeat advocates. An e-commerce business that implements personalized recommendations and frictionless checkout processes, for example, strengthens customer relationships and boosts retention.

By embedding UX principles into digital transformation, businesses lay the groundwork for exceptional experiences, sustained growth, and long-term adaptability. In this context,

UX evolves from a design discipline into the engine of innovation and resilience in an increasingly complex digital landscape.

UX Activities That Support Integration

Practical tools and workshops are essential for successfully integrating UX principles into your organization. These activities help teams align around user-centered goals, foster collaboration, and make UX integration feel natural rather than disruptive. By involving cross-functional groups and creating shared experiences, these methods build empathy, strengthen alignment, and highlight the value of UX-driven approaches.

Design Sprints: Rapid Problem Solving
Design sprints are focused, time-boxed sessions where teams collaborate to solve specific challenges. Over the course of a few days, participants define the problem, brainstorm solutions, prototype ideas, and test them with real users. For instance, a retail company might use a design sprint to prototype a mobile app feature to reduce cart abandonment.

Why It Works: Design sprints foster innovation by encouraging teams to take risks and iterate quickly. They result in actionable prototypes that can be tested and refined, accelerating the design and development process.

Usability Testing Observations: Building Empathy
Inviting team members to observe real users interacting with a product is a powerful way to build empathy and reinforce the importance of user-centered design. Observing users struggle with unclear navigation or celebrating a seamless experience makes the impact of UX tangible for stakeholders.

Why It Works: Usability testing turns abstract concepts into real-world insights. It helps teams prioritize improvements that matter most to users while fostering a shared commitment to quality and usability.

Metrics Alignment Workshops: Connecting UX to Business Goals

Metrics alignment workshops create a direct link between UX metrics and broader business KPIs. Teams collaborate to define success, using UX metrics like task completion times, satisfaction scores, or drop-off rates to measure progress toward business objectives like increased revenue or reduced churn.

Why It Works: These workshops ensure everyone works toward the same goals and understands how their contributions impact the organization's success. They also help business leaders understand the return on UX investments.

Empowering Teams with UX-Driven Approaches

These tools and workshops are more than just exercises, they are frameworks for creating alignment, driving innovation, and demonstrating the tangible impact of UX. When teams collaborate through activities like journey mapping, design sprints, usability testing, and metrics alignment, they gain a shared understanding of user needs and business goals. Incorporating these practices into your organization creates a culture where UX is a natural part of the decision-making process.

A Balanced Approach to Influence

Adopting a product-focused model, where UX, product
management, and engineering operate as equal partners,
is widely regarded as essential for delivering exceptional
user experiences. However, the path to this model varies.
Some organizations favor gradual transitions, while others
advocate for comprehensive transformations. Regardless of
the approach, there has never been a better time to begin this
journey.

Integrating UX principles is not about giving UX outsized
influence; it's about fostering balance. Just as marketing
strengthens brand positioning and operations drive efficiency,
UX ensures that the experiences you deliver align with user
needs and expectations. When UX is treated as an equal partner
in shaping strategy, it enhances your organization's ability to
innovate and grow.

Integrating UX with Cross-Functional Teams

A practical starting point is embedding UX practitioners into
cross-functional teams and aligning their work with product
and engineering goals. Encourage collaboration, shared
ownership of outcomes, and an iterative approach to integrating
UX into broader business strategies. This gradual evolution
minimizes disruption while steadily increasing alignment and
effectiveness. By encouraging user-centered thinking across
your teams, you empower them to design solutions that work
for your customers and your entire organization. A UX-driven
culture is not a one-time shift; it's an ongoing evolution that

equips your business to adapt, grow, and thrive in a competitive marketplace.

As your organization takes steps toward embedding UX more deeply, it is essential to think beyond individual teams or projects. Scaling UX across the entire organization creates the foundation for delivering consistent, impactful experiences at every touchpoint. In the next chapter, we'll explore how to expand UX influence across your business, ensuring that its principles and practices deliver maximum value at scale.

NOTES

Chapter 7: Scaling UX Across the Organization

If a picture is worth 1000 words, a prototype is worth 1000 meetings.

Tom & David Kelley
Creative Brothers—IDEO

For many businesses, UX exists in some capacity, often within a single team or focused on one area of the organization, such as product design or marketing. While this is a great starting point, UX may not yet have a broader influence on other key areas like operations, customer support, or internal tools. Business leadership within organizations with no dedicated UX function or those with UX siloed in one department face a similar challenge: thoughtfully expanding UX influence to align with business goals and strengthen existing processes.

Expanding UX is about strategically testing, measuring, and introducing UX practices in targeted areas to demonstrate value. By scaling gradually, businesses can build confidence in UX's potential while addressing real user needs across customers and employees.

Where to Get Started

For many companies, incorporating UX has been a mixed experience. Some have invested in UX initiatives only to feel like they didn't see the full value, leading to frustration or even resource cuts. This often happens when UX is introduced without a clear strategy, alignment with business goals, or the right expertise in place. The good news is that it's never too late to reset. Start by evaluating where UX fits your organization today: Are efforts siloed or ad hoc? Is there alignment between UX activities and measurable business objectives? From there, focus on small, high-impact projects demonstrating the value of user-centered design. Building from these successes, and pairing them with the right talent and leadership, can help reignite confidence in UX and unlock its potential to drive meaningful, measurable results for your business.

The Gartner Hype Cycle for User Experience

The Gartner Hype Cycle for User Experience, 2024, created by Gartner, a leading global research and advisory firm, provides a framework for understanding how emerging UX technologies progress from innovation to mainstream adoption.[30] By identifying their position on this curve, businesses can strategically plan the integration of new technologies to enhance user experiences and achieve meaningful outcomes.

Five Phases of the Hype Cycle

The Hype Cycle includes five phases. The Innovation Trigger sparks initial interest but often lacks practical applications.

The Peak of Inflated Expectations sees enthusiasm driven by early successes and hype. This is followed by the Trough of Disillusionment, where setbacks temper excitement. The Slope of Enlightenment brings clarity as benefits and practical applications emerge, leading to the Plateau of Productivity, where technologies achieve widespread adoption and measurable value.

Highlighted technologies include Generative AI-enabled applications, AI-augmented software engineering, and spatial computing, which enhance interaction and immersion. The Digital Twin of a Customer improves predictive insights, while Total Experience (TX) integrates customer, employee, and user experiences to create seamless interactions across touchpoints.

The Hype Cycle equips businesses to navigate the evolving UX landscape confidently. By aligning adoption with strategic goals, monitoring advancements, and investing in necessary resources, organizations can position themselves for innovation and long-term growth. It serves as a roadmap, helping companies transition from their current state to a future where UX drives measurable success.

Creating Immediate Impact

Introducing or expanding UX within an organization requires focusing on areas where it can deliver quick, measurable results. Whether UX is entirely new or siloed within one team, targeted initiatives can demonstrate its value and build momentum for broader adoption. If UX already exists within product design, consider extending its influence to other areas, such as employee-facing tools or operational workflows. For instance, if customer service teams struggle with outdated systems, UX improvements can streamline processes, reduce errors, and improve response times. Similarly, enhancing internal dashboards for sales teams can free up time for customer interactions, directly impacting revenue generation.

Encourage collaboration between UX professionals and teams outside their traditional scope. For example, if UX primarily focuses on customer-facing touchpoints, involve them in discussions about internal operational tools or workflow optimization. These cross-functional partnerships often uncover overlooked opportunities to drive efficiency, innovation, and user satisfaction. Another effective way to expand UX's influence is to introduce a UX researcher to an existing project to conduct validation testing with users. For example, if a new feature is in development, having a UX researcher gather real-world user feedback can ensure the feature meets user needs and aligns with business objectives. This targeted intervention helps reduce risks, refine solutions, and accelerate time-to-market by addressing potential issues before full deployment.

Lay the foundation for integrating user-centered thinking across the business.

Define success metrics for each initiative, whether reducing task completion times, improving customer satisfaction scores, or increasing conversion rates. For example, if a project focuses on refining an onboarding flow, success might be measured by a decrease in drop-off rates and a corresponding boost in user retention. Aligning UX efforts with tangible business outcomes ensures its value is clear and measurable. Organizations can achieve immediate, demonstrable results by identifying high-impact opportunities, fostering cross-functional collaboration, and introducing targeted UX research. These initiatives highlight UX's potential and lay the foundation for integrating user-centered thinking across the business.

Measuring Impact to Build Momentum

Demonstrating the measurable impact of UX initiatives is critical for gaining organizational buy-in, especially when expanding UX beyond its current scope or introducing it to new areas. Start by tracking metrics that are directly tied to UX efforts. For instance, if a redesigned employee tool reduces task completion times by 30%, calculate the corresponding cost savings or productivity gains to showcase the tangible benefits.

In addition to quantitative data, user feedback from customers and employees should be gathered to understand how the changes have improved their experiences. Combining stories with metrics provides a compelling narrative that underscores the value of UX. This qualitative input adds depth to the numbers, illustrating how improved design translates to better engagement and satisfaction.

Sharing Results

Finally, share these results widely across leadership and relevant teams. Highlight how UX improvements align with broader business goals, such as smoother internal workflows enabling faster customer response times or enhanced satisfaction scores. Connecting UX outcomes to strategic objectives builds momentum for future initiatives and reinforces the importance of user-centered design in driving business success.

IBM's Thoughtful Execution in UX Expansion

As we established in Chapter 1, IBM's Enterprise Design Thinking initiative delivered remarkable ROI, including substantial improvements in employee productivity and customer satisfaction. However, behind those results lies a case study on how to execute a company-wide UX transformation effectively. By focusing on thoughtful execution and strategic alignment, IBM's approach offers valuable lessons for any organization seeking to integrate UX principles across its operations.[31]

IBM recognized that fragmented user experiences and inefficiencies limit customer satisfaction and internal productivity. The company launched Enterprise Design Thinking to address these challenges, a comprehensive initiative to embed UX principles throughout the organization. This was not a top-down directive but a carefully structured effort to create cultural alignment and build capabilities at scale.

A cornerstone of IBM's approach was training more than 100,000 employees in user-centered design. This training extended beyond traditional design roles to include developers, business leaders, and key stakeholders. The goal was to establish a shared understanding of UX principles and foster a common language for collaboration across the organization. This widespread training helped break down silos and align teams around user needs, ensuring UX considerations became part of every project.

Centralized Leadership

To maintain consistency and direction, IBM established a centralized UX leadership team. This team was responsible for providing clear frameworks, best practices, and strategic oversight to ensure that UX efforts aligned with business objectives and upheld the principles of Enterprise Design Thinking. Cross-functional teams of designers, developers, and business stakeholders tackled specific challenges, using iterative design methods to drive collaboration and innovation.

IBM's disciplined execution yielded significant operational improvements. Internal tools were redesigned to streamline workflows, enabling employees to work more efficiently and collaborate seamlessly. Development cycles were reduced by 75%, reflecting the power of iterative processes and better alignment between teams. For customers, the consistent application of UX principles resulted in more intuitive interfaces and enhanced satisfaction, strengthening trust in the IBM brand.

This case underscores the importance of a deliberate and structured approach to scaling UX. By prioritizing employee training, fostering cross-functional collaboration, and leveraging centralized leadership, IBM created a culture where user-centered thinking thrives. The measurable outcomes:

faster delivery times, more intuitive tools, and enhanced customer satisfaction, were not the result of chance but careful planning and execution. IBM's journey highlights the potential for UX to improve individual touchpoints and transform how an organization operates as a whole.

Barriers to UX Success and How to Overcome Them

While the value of UX is widely recognized, many organizations face obstacles that prevent UX initiatives from achieving their full potential. By addressing these barriers and implementing targeted solutions, businesses can unlock the full power of UX to drive user satisfaction and business success.

1. Lack of Leadership Buy-In

One of the most significant barriers to UX success is the absence of strong advocacy from leadership. Without executive support, UX often lacks the visibility and authority to influence strategy and achieve impact.

Solution: Educate leadership on the strategic value of UX by connecting it to measurable business outcomes like increased revenue or customer retention. Share case studies and real-world data that demonstrate how UX can drive growth. Invite executives to participate in UX workshops or usability testing sessions to experience its value firsthand.

2. Misaligned Goals

When UX goals are not aligned with broader business objectives, its contributions can seem disconnected or superficial. For example, prioritizing aesthetic updates over usability improvements without addressing KPIs like retention or conversion rates can diminish the perceived impact of UX.

Solution: Align UX metrics with organizational goals. For instance, link usability improvements to measurable KPIs such as customer retention, revenue growth, or reduced churn. Conduct metrics alignment workshops to ensure all teams understand how UX supports broader objectives, fostering a shared commitment to measurable success.

3. Resistance to Change

Shifting to a user-centered approach often requires rethinking workflows and processes, which can be met with resistance, especially in organizations with entrenched habits or hierarchical structures.

Solution: Address resistance through gradual change management. Start with pilot projects that demonstrate the value of UX in a specific, low-risk context. Involve resistant team members in UX activities such as journey mapping or usability testing to build empathy and understanding. Highlight quick wins to show how UX improvements can drive tangible, immediate results.

4. Poor Integration with Development Teams

A lack of collaboration between UX and engineering teams often results in misaligned priorities, suboptimal implementation, and poor user experiences. This issue is exacerbated when development teams prioritize speed and volume of output over quality and usability. Additionally, insufficient time allocated for UX activities before development begins can lead to rushed designs that fail to address user needs effectively.

Solution: Foster strong partnerships between UX and engineering by establishing shared goals and regular touchpoints, such as joint design reviews and collaborative sprint planning. Advocate for measurable quality improvements, like reduced error rates or faster task completion times, to ensure development efforts align with user-centered objectives. Before development begins, allocate adequate time for UX activities, such as research, wireframing, and usability testing. This allows teams to identify and address potential issues early, avoiding costly rework later. Create a shared understanding between UX and engineering of how thoughtful design contributes to product quality and development efficiency. By balancing speed with user-centered outcomes, teams can deliver products that meet business objectives and user needs.

5. Inadequate Focus on Research

Skipping or deprioritizing user research is a common challenge in many organizations. This approach often results in designs based on assumptions rather than evidence, frequently leading to solutions that fail to address user needs effectively, reducing user satisfaction and adoption rates. One key issue is the insufficient time allocated for research at the beginning of projects, which limits teams' ability to gather meaningful insights and validate design decisions.

Solution: Make user research a foundational component of UX practices, ensuring it is prioritized and allocated adequate time in project timelines. Use qualitative methods like interviews, usability testing, and focus groups to gain rich insights into user behaviors and pain points. Combine these with quantitative data from analytics to create a comprehensive picture of user needs.

6. Poor Problem Definition and Lack of Framing

Starting a project without clearly defining the problem, framing the scope, or setting realistic expectations often leads to wasted efforts and misaligned outcomes. Teams may create solutions for the wrong problems or fail to deliver results that meet organizational needs.

Solution: Begin every project with a clear problem definition and a shared understanding of goals. Use framing exercises like stakeholder interviews, user journey mapping, and value proposition canvases to align teams on the objectives and constraints. Ensure expectations are realistic and well-communicated to avoid scope creep or misaligned deliverables. Regularly revisit these definitions during the project to maintain alignment and focus.

Overcoming Silos and Encouraging Collaboration

One of the most significant challenges in expanding UX influence is breaking down silos. When UX is confined to a single department, its potential impact is limited. To unlock the full value of UX, leadership must foster collaboration and ensure UX insights benefit the entire organization. Cross-functional partnerships are a key strategy. UX teams should be encouraged to collaborate with sales, operations, marketing, and IT. For instance, a UX professional might work with IT to improve internal tools and streamline employee workflows while supporting marketing in optimizing lead-generation workflows to enhance customer acquisition efforts.

> *Foster collaboration and ensure UX insights benefit the entire organization.*

Another essential step is creating shared goals. By aligning departments around common objectives, such as improving customer satisfaction or reducing operational inefficiencies, leadership can ensure that UX initiatives are considered integral to broader business success rather than isolated projects. This shared focus helps teams see UX as a strategic partner in achieving measurable outcomes.

Finally, organizations should leverage existing UX expertise. If UX talent is already in place, give them visibility into other areas of the business where their skills can make a meaningful impact. This increases their influence and allows the organization to benefit from user-centered thinking across broader initiatives.

The Role of Expertise and Leadership in UX

Design thinking is often a starting point for organizations introducing or expanding UX. This human-centered problem-solving framework emphasizes empathy, creativity, and iterative prototyping. It fosters collaboration and helps teams generate user-focused solutions. It's particularly effective for building awareness, encouraging innovation, and promoting a user-centered mindset. However, design thinking alone is not sufficient to unlock the full potential of UX, especially at the enterprise level.

Supporting Design Thinking with Experts

While design thinking excels at ideation and aligning cross-functional teams, it has limitations. It often lacks the depth needed to address complex, system-wide challenges or validate solutions through rigorous research. Furthermore, its focus on early-stage ideation doesn't extend into execution, leaving gaps in scalability, usability, and implementation. To deliver measurable results, organizations must pair design thinking with the expertise of trained UX professionals: designers, researchers, and leaders who bring rigor, depth, and strategic alignment to the process.

UX professionals ensure that solutions are grounded in evidence through robust user research, refined through iterative testing, and aligned with business goals by skilled leadership. At scale, UX leadership is critical in advocating for resources, aligning efforts with strategic priorities, and ensuring collaboration across teams. By balancing the creativity of design thinking with the structure and expertise of a mature UX practice, organizations can move from ideation to execution, delivering solutions that drive user satisfaction and measurable business success.

Why Expertise Matters

UX is a specialized discipline that goes far beyond brainstorming sessions or empathy exercises. Trained designers and researchers bring a depth of expertise and proven methodologies that are essential for creating impactful solutions. Understanding user needs requires more than surface-level surveys. UX researchers employ advanced techniques such as ethnographic studies, usability testing, and data analysis to uncover deep insights into user behaviors and pain points. This research forms the foundation for designing experiences that truly resonate.

Designing seamless and intuitive experiences is not merely about aesthetics. Skilled UX designers balance user needs, technical constraints, and business goals to create solutions that are not only functional but also impactful. They ensure the designs align with strategic objectives while delivering a frictionless user experience.

Effective iteration is another hallmark of UX expertise. Professionals with UX training refine prototypes and designs based on user feedback, ensuring potential issues are identified and addressed early in the process. This iterative approach minimizes costly mistakes and maximizes the likelihood of success. Without this level of expertise, companies risk making superficial changes that fail to address core user pain points or meet business objectives. Trained UX professionals bring the necessary knowledge and tools to design solutions that deliver measurable value, ensuring that user-centered thinking drives meaningful results.

Why UX Leadership Matters

As UX practices expand within an organization, strong leadership becomes essential. UX leaders are pivotal in translating high-level business objectives into actionable UX strategies, ensuring design teams align with broader organizational goals. They also advocate for the user perspective, ensuring customer and employee needs are consistently represented in decision-making processes.

At scale, the importance of UX leadership becomes even more pronounced. Leaders in this space bridge the gap between business priorities and design activities, guiding teams to focus on initiatives that align with the company's strategic objectives. They foster cross-functional collaboration, ensuring that UX efforts integrate seamlessly with the work of other departments like marketing, engineering, and operations. This alignment helps create cohesive and impactful user experiences.
As UX practices grow, leadership ensures consistency in methodologies, quality, and outcomes. By establishing scalable frameworks, UX leaders create a foundation for sustained value delivery, enabling the organization to maintain high standards even as the scope of UX efforts expands.

Without dedicated leadership, UX efforts risk becoming fragmented or misaligned with business priorities. UX leaders provide the vision, strategy, and structure necessary to unify these efforts and maximize their impact. They ensure that UX practices enhance individual touchpoints and drive business impact across the organization.

Budgeting for UX: Investing Strategically

Budgeting for UX doesn't have to be overwhelming or resource-intensive, especially for organizations just starting. A strategic approach begins with small, focused investments in areas where UX can deliver quick and measurable results. For example, piloting UX improvements in a specific product or workflow allows businesses to demonstrate value without committing significant upfront costs. Resources should also be allocated for essential tools like analytics platforms, design and prototyping software, or usability testing systems. For organizations not ready to hire full-time UX professionals, outsourcing to consultancies or freelancers provides access to expertise on a project basis, ensuring flexibility and high-quality execution.

UX budgeting can follow different models depending on the organization's structure and goals. For businesses with established product roadmaps, budgets can focus on supporting delivery, ensuring UX initiatives align with key milestones and strategic priorities. Alternatively, companies with persistent teams can allocate annual budgets to sustain ongoing UX efforts, fostering a culture of continuous improvement. For more dynamic needs, a project-focused approach allows teams to adjust resources as required, providing the agility to address emerging opportunities or challenges effectively.

Preparing to Scale UX

Scaling UX across your organization is less about sweeping transformations and more about finding the right opportunities to expand its influence thoughtfully. By starting small, aligning efforts with strategic goals, and demonstrating measurable outcomes, businesses can integrate UX to deliver value across teams and departments. This approach fosters confidence in UX and builds the foundation for a culture of user-centered thinking that benefits both customers and employees.

Setting up for Success

This is an opportunity to recalibrate for organizations that have struggled to see the full value of their UX investments. By focusing on clarity, alignment, and collaboration, you can ensure that UX efforts directly support business objectives and drive tangible outcomes. Whether improving internal tools, enhancing customer experiences, or aligning cross-functional teams, the goal is to create a system where UX contributes meaningfully to your organization's success.

In the next chapter, we'll move from strategy to action. You'll learn practical steps for integrating UX into your organization, from setting clear goals to measuring impact and fostering collaboration. These steps will equip you to take the insights from this chapter and turn them into actionable, results-driven initiatives that demonstrate the power of great UX.

NOTES

Chapter 8:
Practical Steps for Success

Testing with one user early in the project is better than testing with 50 near the end.

Steve Krug
Author "Don't Make Me Think"

By now, it's clear that UX has the potential to transform your business: enhancing customer loyalty, improving employee workflows, and driving measurable outcomes. The next step is action. This chapter provides a practical guide to integrating UX into your organization, whether starting with a single initiative or scaling it across multiple teams. By focusing on alignment, tools, and clear communication, you can ensure UX delivers meaningful value to your business.

Building a Thriving UX Community of Practice

A community of practice is more than just a gathering of professionals. It's a dynamic network where individuals deepen their expertise, share insights, and collaboratively advance their craft. In the context of UX, this kind of community fosters a culture of collaboration, continuous learning, and excellence. Bringing together designers, researchers, and strategists provides a structured framework for sharing successes, addressing challenges, and ensuring that UX efforts align with organizational goals.

Growing Together

The value of a strong UX community of practice extends beyond individual growth to benefit the entire organization. Through knowledge sharing, members exchange ideas, tools, and techniques, collectively raising the quality of UX work. Regular feedback loops encourage continuous improvement, helping teams refine their approaches and fostering a culture of learning and adaptation. This collaboration also ensures alignment and consistency, as shared principles and goals guide UX efforts across projects, keeping them cohesive and aligned with organizational strategy.

A community of practice also offers vital support and collaboration, creating a space where challenges are tackled collectively and successes are celebrated. Increased visibility through regular touchpoints with leadership and cross-functional teams showcases the impact of UX, building advocacy for the discipline and reinforcing its importance within the organization. By nurturing this professional ecosystem, a UX community of practice becomes a key driver of growth, innovation, and alignment within the business.

To build a strong UX community of practice, it's essential to establish consistent, structured opportunities for collaboration, feedback, and transparency. Three foundational touchpoints create the basis for a thriving community:

1. Collaborative Cross-Functional Touchpoints (Project Level)

At the project level, collaborative touchpoints bring together UX professionals, product managers, engineers, and other stakeholders. These sessions foster creative problem-solving by enabling diverse teams to address challenges collectively. They enhance communication, ensuring alignment on project goals and progress and establishing accountability among team members.

2. Design Reviews (UX Practice Level)

Recurring design reviews are essential for maintaining high standards and promoting continuous improvement in the UX practice. Designers share work in progress, receive constructive feedback, and refine their ideas. These reviews drive execution excellence, encourage sharing insights and best practices, and foster accountability by ensuring alignment with user needs and organizational priorities.

3. Design Showcases (Leadership Engagement)

Design showcases provide a platform for UX teams to share their work with cross-functional leadership, creating transparency and building advocacy for UX initiatives. These events highlight the impact of design efforts, reinforce the value of user-centered approaches, and align UX strategies with broader business goals. By engaging leadership, design showcases help secure ongoing support for UX investments.

A thriving community of practice strengthens the UX discipline at every level. Collaborative touchpoints at the project level ensure alignment and creativity. Design reviews within the practice uphold quality and consistency. Design showcases engage leadership, fostering trust and demonstrating value. Together, these elements create a cohesive system that supports individual growth, organizational alignment, and the delivery of exceptional user experiences.

Equipping Teams with the Right Tools

For UX teams to thrive, they need access to the right resources. Designers and researchers rely on specialized tools that streamline workflows, foster collaboration and uncover actionable insights. Providing these tools enhances efficiency and quality and signals your organization's commitment to supporting user-centered design efforts.

Design and Research Tools

Prototyping and design tools like Figma, Sketch, and Adobe XD enable designers to create interactive prototypes that bring ideas to life. These tools also allow teams to visualize and iterate on solutions before committing to development, facilitating rapid experimentation and reducing the risk of costly missteps. Usability testing platforms such as UserTesting, Lookback, or Maze empower researchers to gather real-time feedback from users, helping them identify pain points, test assumptions, and validate design decisions. This ensures that final solutions align with user needs and expectations.

Analytics platforms, including Hotjar, Google Analytics, and FullStory, provide critical insights into how users interact with products. Heatmaps, session recordings, and user flow reports highlight areas for improvement, enabling data-driven decisions and prioritization of high-impact efforts. For organizing and synthesizing research findings, tools like Dovetail and Airtable streamline processes and help share insights with stakeholders. Collaboration platforms such as Miro, Slack, and Microsoft Teams offer distributed and cross-functional teams a shared space to effectively brainstorm, align, and track progress.

UX practitioners excel when supported by the right resources.

Accessibility testing tools, such as Axe, WAVE, and Stark, ensure that designs meet the needs of all users, including those with disabilities. These tools help organizations comply with standards like WCAG and demonstrate a commitment to inclusivity, broadening the reach and impact of their digital experiences.

UX practitioners excel when supported by the right resources. These tools amplify your team's capability to solve problems faster, create more impactful designs, and deliver consistent results. By equipping your UX team with the right tools, you optimize their potential and prepare them to produce their most effective and innovative work.

Setting Practioner Goals

As we've clearly established, to deliver real value, UX activities must align with broader business objectives. For UX efforts to have a tangible impact, it's essential to set measurable outcomes for individual contributors, such as designers and researchers, that can be tracked independently of other teams.

Clear, measurable goals help focus UX efforts while fostering accountability and motivation. Here are some examples of what these goals might look like for individual contributors:

For a UX Researcher:
Goal: Conduct usability testing on five critical workflows within the next quarter and deliver actionable recommendations to improve task completion rates by 10%. This goal ensures the researcher focuses on uncovering insights and presenting findings that directly improve the product experience.

For a UX Designer:
Goal: Develop three fully interactive prototypes for proposed feature enhancements, ensuring that each prototype adheres to established accessibility guidelines and is ready for user feedback within four weeks. This goal emphasizes the designer's role in creating detailed, user-centered prototypes.

For a UX Content Strategist:
Goal: Based on a review of customer feedback and analytics data, the goal is to rewrite the error message copy across the platform and reduce support tickets by 15% within two months. This focuses on making content improvements that can be implemented quickly and yield measurable results.

For a UX Leader:

Goal: Within the next six months, establish a scalable UX strategy that aligns with the company's business objectives, including implementing a design system across all product teams to ensure consistency and efficiency. This includes facilitating three cross-functional workshops to align stakeholders and defining metrics to measure the success of UX initiatives, such as a 15% reduction in time-to-market for key product features.

By aligning these individual goals with broader business objectives like improving usability, increasing accessibility, or reducing support costs, UX professionals can contribute to measurable outcomes while maintaining autonomy over their work. Regular check-ins with UX leaders and stakeholders ensure these efforts align with company priorities and create a clear connection between UX contributions and business impact.

Clear, measurable goals help focus UX efforts while fostering accountability.

In addition to aligning with business objectives, UX practitioners should set goals that foster team growth and personal development. Focusing on these areas they help create a thriving and innovative UX practice. Growing the team involves mentoring junior colleagues, leading workshops, or developing design resources that enhance collective expertise and cohesion. Practitioners who set goals related to teaching or team-building activities contribute to a supportive and dynamic environment, ensuring that the UX practice evolves as a whole. These efforts strengthen collaboration, encourage knowledge sharing, and build a more resilient team.

Equally important is growing themselves. Continuous learning enables UX professionals to stay ahead of industry trends and technological advancements. Practitioners can set personal development goals, such as mastering new tools, attending conferences, or completing specialized training programs. These activities expand individual capabilities and enrich the broader UX practice by introducing fresh perspectives and skills. These goals ensure the UX function remains innovative, adaptive, and aligned with the organization's evolving needs.

By aligning their efforts with organizational goals, contributing to team development, and pursuing personal growth, UX practitioners maximize their impact at every level: creating value for themselves, their teams, and the business as a whole.

Start Driving Impact and Alignment

Certain design activities do more than solve immediate challenges. They bring teams together, align efforts, and clarify long-term strategies. These high-value activities combine collaboration, creativity, and user-centered approaches to address complex problems while ensuring that teams work toward shared goals. Here's how these activities can drive alignment and value within your organization:

Business Blueprint
What It Is: A comprehensive map of how your organization's processes, systems, and teams contribute to the overall experience, identifying areas for optimization.

Example:

A healthcare provider creating a business blueprint might document how patient scheduling, billing, and follow-up communications flow through various departments. The blueprint helps pinpoint inefficiencies, like redundant data entry in scheduling, and creates a roadmap for streamlining operations while improving the patient experience.

System Inventory Mapping

What It Is: System inventory mapping catalogs all internal and third-party systems, tools, and processes that impact the user experience. This exercise clarifies how systems interact, highlights redundancies, and identifies gaps creating inefficiencies. Including third-party solutions is essential, as they often play a significant role in workflows but can introduce inconsistencies if not well-integrated.

Example:

A financial services firm discovered employees were using five separate tools, both internal and third-party, to manage customer accounts. This redundancy caused errors, wasted time, and frustrated employees. By consolidating tools into a single platform, the company streamlined operations saved time, and improved customer service through faster, more accurate responses.

Vision Workshops

What It Is: Collaborative sessions to define an aspirational "north star" for the experience, creating alignment around shared goals.

Example:

During a vision workshop, a travel company might brainstorm what a seamless booking experience could look like in five years. The team defines aspirational goals, like integrating voice-activated search or personalized travel recommendations, to guide future UX initiatives.

Journey Mapping

What It Is: A visualization of the end-to-end experience of a customer or employee, highlighting touchpoints, pain points, and opportunities for improvement.

Example:

A journey mapping session for a retail company could explore the customer's path from online product discovery to in-store pickup. Marketing, logistics, and customer support teams work together to uncover friction points, such as a lack of real-time inventory updates, and align on solutions, like integrating a live inventory system to improve customer trust and satisfaction.

User Discovery Interviews

What It Is: In-depth conversations with users to uncover their needs, behaviors, and pain points, providing actionable insights for design and strategy.

Example:

A SaaS company launching a new collaboration tool might conduct discovery interviews with existing customers to learn about their workflows and frustrations with current tools. Insights gathered could guide feature prioritization, ensuring the product addresses real user needs.

User Success Metrics

What It Is: Defining clear goals for user success, creating systems to measure these outcomes, and understanding how they correlate with business KPIs like revenue, retention, or efficiency.

Example:

An e-commerce company might define user success in its checkout process as completing a purchase without errors or confusion. To measure this, the company could track metrics like task completion rates, time to check out, and error frequency. By analyzing these metrics alongside business KPIs, such as cart abandonment rates or average order value, the company can identify causal relationships. For instance, improving the checkout experience could lead to higher conversion rates and increased revenue.

Why These Activities Matter

These high-value activities produce actionable results, foster collaboration, align teams, and ensure that UX initiatives are strategically focused. They provide a clear path forward, helping organizations balance creativity with practicality while driving measurable business outcomes. Incorporating practices like journey mapping, discovery interviews, and prototyping ensures that your UX efforts remain user-centered, impactful, and aligned with your company's broader goals.

Three Key UX Principles

Defining and delivering solutions requires a thoughtful approach guided by principles that shape how we work and ensure success. These principles: cross-functional collaboration, iterative progress, and user-centered design, are the foundation of a strong UX practice.

1. UX Is Cross-Functional

Effective UX doesn't happen in isolation. It thrives through collaboration with strategic partners across the organization, including product managers, engineers, marketers, and business leaders. This cross-functional approach ensures that diverse perspectives are integrated into the process, creating solutions that balance user needs with technical feasibility and business priorities. Collaboration builds alignment and shared ownership. When UX is a hub for cross-functional input, teams work more cohesively, communication improves, and solutions become more comprehensive and impactful.

2. UX Works Iteratively

Good design isn't created in a single attempt. The UX process is iterative by nature: proposing ideas, testing them, learning from feedback, and refining the solution. Each iteration brings the design closer to the optimal outcome by addressing challenges and incorporating new insights. Iteration reduces risk and drives improvement. By testing and refining solutions early and often, we avoid costly missteps, ensure quality, and adapt to evolving needs. This mindset fosters continuous learning and keeps teams focused on delivering the best possible results.

3. UX Is User-Centered

Being user-centered is more than a principle. It's a commitment to actively engaging users throughout the design process. This means conducting research to understand their needs, testing ideas with real users, and folding their insights into every stage of development. User engagement ensures relevance and resonance. Solutions grounded in real-world feedback are more likely to meet user expectations, drive satisfaction, and deliver measurable business value. A user-centered approach not only reduces friction but also builds trust and loyalty.

Why These Principles Matter

Together, these principles define how UX delivers value: fostering collaboration, embracing iteration, and designing for users. They ensure that UX efforts lead to thoughtful, effective solutions aligned with strategic objectives. By adhering to these principles, UX solves problems and strengthens teams, builds trust with users, and drives success for the organization.

Having Open Conversations About Impact

As UX becomes more integrated into the organization, maintaining open and regular dialogue with UX leadership is essential. These conversations help ensure UX initiatives align with business priorities and deliver measurable results. Regular touchpoints, such as one-on-one meetings and annual performance reviews, play a crucial role within the UX community of practice.

The Role of One-on-Ones and Performance Reviews

One-on-one meetings provide recurring opportunities for UX practitioners and their leaders to discuss progress, address challenges, and align on short-term and long-term goals. These conversations foster a sense of support, encourage professional development, and create space to address immediate concerns or needs. By maintaining open lines of communication, leaders can ensure that practitioners feel valued and empowered in their roles.

Annual performance reviews take a broader view, assessing the impact of a practitioner's work over the year and identifying growth opportunities. These structured discussions should celebrate successes, provide actionable feedback, and set clear, measurable goals for the year ahead. Performance reviews not only reinforce accountability but also help practitioners see how their contributions align with team and organizational objectives.

Together, one-on-one meetings and performance reviews form a framework for growth and alignment, ensuring that individual efforts contribute meaningfully to the success of the UX practice and the organization.

Fostering Transparency and Continuous Improvement

Fostering transparency and continuous improvement is essential for ensuring the long-term success and impact of UX initiatives. Open conversations about the outcomes of UX projects create opportunities to celebrate successes, address challenges, and collaboratively refine strategies. Celebrating successes highlights the measurable outcomes of UX efforts, such as increased user satisfaction or reduced task completion times.

By sharing these achievements, teams can demonstrate the value of their work and build advocacy for UX across the organization. At the same time, these discussions should be a platform for identifying challenges, surfacing roadblocks, or addressing areas where UX efforts may require additional support or alignment with broader business objectives.

Finally, these conversations allow teams to refine their approaches collaboratively. By brainstorming solutions and adjusting strategies based on feedback, UX practitioners and stakeholders can maximize the impact of future initiatives. Transparent and open dialogue builds trust across the organization and reinforces a culture of continuous improvement, ensuring that UX remains a vital contributor to organizational success.

The Foundation for UX Success

Creating a thriving UX practice isn't just about having the right tools or processes. It's about embedding UX into the very fabric of your organization. From fostering a community of practice that emphasizes collaboration and learning to ensure alignment with broader business goals, this chapter has outlined how to implement UX in ways that deliver tangible results. Whether starting with small, targeted initiatives or scaling UX across your organization, these steps create a foundation for sustainable growth and meaningful impact.

The journey doesn't stop here. As the world evolves, so do the tools, techniques, and technologies that shape user experiences. Staying ahead requires more than just a commitment to UX. It demands an understanding of emerging trends and the ability to adapt your approach to leverage new opportunities. In the next chapter, we'll explore how the latest trends and technologies are redefining UX and providing businesses with innovative ways to create value. By embracing these advancements, you'll position your organization to thrive in a rapidly changing landscape.

NOTES

NOTES

Chapter 9:
Trends and Technologies

*Technology should bring more to our lives
than the improved performance of tasks:
it should be richness and enjoyment.*

Don Norman
Author and UX Pioneer

The rapid pace of technological innovation is transforming how businesses operate and interact with their users, raising the stakes for delivering seamless, impactful experiences. Emerging technologies like artificial intelligence (AI), hyper-personalization, augmented reality (AR), and blockchain are no longer optional considerations. They are reshaping customer expectations and redefining exceptional UX in the digital age. These advancements are not just tools; they represent a paradigm shift in how companies connect with their customers and empower their employees.

These technologies present immense opportunities. AI can analyze vast amounts of user data to anticipate needs and deliver hyper-personalized interactions. At the same time, AR and virtual reality (VR) offer immersive experiences that bridge the gap between physical and digital worlds. Blockchain revolutionizes trust in digital transactions by providing transparent and secure data management. At the same time, voice interfaces, generative AI, and predictive analytics are creating new dimensions of engagement that demand thoughtful UX integration.

Technological innovation is transforming how businesses operate and interact with their users

However, these innovations come with challenges. Businesses must navigate rapid adoption cycles, integrate new tools into existing systems, and ensure these technologies meet user needs without sacrificing privacy, trust, or accessibility. Successful companies will go beyond adopting emerging tech for its own sake; they will thoughtfully align these tools with user-centered design principles to create meaningful, lasting impact.

This chapter explores how these technologies are reshaping the UX landscape, providing businesses with both opportunities and imperatives. By understanding the potential and pitfalls of these innovations, organizations can position themselves as leaders in an increasingly digital and interconnected world. Now is the time to embrace the future, not just by adopting new technologies but by embedding them into experiences that align with user needs and drive business value.

Leveraging New Technologies to Transform Experiences

Emerging technologies enable businesses to design smarter, faster, and more engaging interactions. Here's how they're reshaping the UX landscape:

Artificial Intelligence

AI is reshaping the UX by enabling hyper-personalization and predictive capabilities that were unimaginable a decade ago. Businesses are leveraging AI to enhance customer and employee experiences. AI-powered chatbots, for instance, can resolve common issues instantly, providing quick and efficient support that reduces user frustration. By handling routine inquiries, these chatbots improve the customer journey and alleviate pressure on customer service teams.

Machine learning algorithms further analyze user behavior in real-time to tailor experiences. For example, e-commerce platforms can recommend products based on browsing history, preferences, and purchase patterns, significantly boosting engagement and conversion rates. On the employee side, AI automates repetitive tasks, such as data entry or report generation, allowing teams to focus on high-impact, strategic priorities.

The benefits are not just theoretical. AI-powered chatbots have reduced support call volumes by 40% for telecom companies, saving millions in operational costs while improving customer satisfaction.[32] As AI continues to evolve, its potential to create seamless, user-centered experiences and optimize business operations will only grow.

Hyper-Personalization

In today's digital landscape, users expect more than generic interactions; they demand experiences tailored to their unique preferences and behaviors. Hyper-personalization leverages advanced data analytics and AI to deliver customized experiences, enhancing user satisfaction and business success.

Retailers, for instance, have significantly benefited from implementing hyper-personalized strategies. Businesses can offer product recommendations that closely align with each user's interests by analyzing individual browsing histories, purchase patterns, and demographic information. This personalized approach has led to substantial increases in engagement and conversion rates. According to Zippia, companies that excel at personalization generate 40% more revenue from those activities than average players.[33]

Thoughtfully align these tools with user-centered design principles to create meaningful, lasting impact

Beyond retail, hyper-personalization is making strides in various sectors. In healthcare, personalized patient portals utilize medical histories and real-time health data to provide tailored health insights and recommendations, improving patient outcomes and satisfaction. In the financial industry, customized dashboards offer clients personalized views of their accounts and investment opportunities, fostering greater user engagement and trust.[34]

The adoption of hyper-personalization is on the rise as businesses recognize its potential to forge deeper connections with users. As technologies like AI and machine learning continue to evolve, the capacity to deliver even more precise and dynamic experiences will become a key differentiator in the competitive digital marketplace.

Voice Interfaces and Conversational AI

Voice-enabled systems are transforming user interactions across various industries, offering intuitive, hands-free experiences that enhance accessibility and convenience. Virtual assistants like Amazon Alexa, Google Assistant, and Apple's Siri have become integral to daily life, enabling users to perform tasks ranging from controlling smart home devices to managing schedules through simple voice commands. Businesses increasingly integrate voice interfaces into customer service and operations to streamline processes and improve user engagement.

In retail, voice interfaces reshape how consumers interact with brands by enabling seamless voice-activated searches and transactions. Healthcare providers are adopting voice technology to improve patient interactions and simplify administrative tasks, such as retrieving patient information or documenting clinical notes, which can be done hands-free. These advancements improve efficiency while ensuring that critical workflows remain uninterrupted.

Despite the clear advantages, implementing voice interfaces requires addressing challenges like speech recognition accuracy across diverse user-profiles and maintaining robust privacy and security measures. By tackling these issues and further embedding this technology into user interactions, businesses can deliver effective and trustworthy voice-enabled experiences.

Augmented Reality (AR) and Virtual Reality (VR)

AR and VR are revolutionizing how businesses interact with customers and train employees. They seamlessly merge physical and digital interactions to create immersive experiences.

In the retail sector, AR enables customers to visualize products within their own environments, enhancing confidence in purchasing decisions. A notable example is IKEA's AR application, IKEA Place, which allows users to place furniture in their homes virtually using a smartphone.[35] This functionality helps customers assess how items fit with their existing décor and spatial constraints, reducing uncertainty and hesitation associated with purchasing large items like furniture. By providing this interactive experience, IKEA enhances customer engagement and satisfaction.

Beyond customer engagement, AR and VR technologies are transforming employee training programs. Companies leverage these immersive tools to create realistic simulations that enhance learning outcomes. For instance, Walmart utilizes VR to prepare employees for high-pressure situations like Black Friday sales.[36] Through virtual simulations, associates can practice managing long lines, addressing customer inquiries, and prioritizing tasks in a controlled, risk-free environment. This approach accelerates proficiency, improves operational efficiency, and boosts employee confidence.

Integrating AR and VR bridges the gap between physical and digital interactions, offering businesses innovative ways to engage with customers and employees. By adopting these technologies, companies can provide personalized, immersive experiences that meet the evolving expectations of modern consumers and create effective, engaging training programs that enhance workforce capabilities.

Blockchain Technology

Blockchain technology is revolutionizing various industries by enhancing transparency, security, and efficiency. Beyond its origins in cryptocurrency, blockchain's decentralized and tamper-proof ledger system enables businesses to build trust and streamline operations. In supply chain management, blockchain facilitates real-time tracking of products from origin to consumer, ensuring authenticity and reducing fraud. For instance, Walmart has implemented a blockchain-based system to trace the origin of produce, significantly reducing the time required to track items from days to seconds.[37]

By leveraging blockchain, businesses can create inherently transparent and secure systems, fostering trust among consumers and partners. As adoption grows, blockchain's potential to address transparency, accountability, and security challenges continues to expand across industries.

Internet of Things (IoT)

The Internet of Things (IoT) is revolutionizing industries by connecting devices to the Internet, enabling real-time monitoring, automation, and data-driven decision-making. IoT spans many applications, from healthcare and logistics to smart homes and industrial operations, providing unprecedented opportunities to enhance workflows, operational efficiency, and user engagement.

IoT-enabled devices are reshaping healthcare delivery by continuously monitoring and fostering preventive care. Wearables like the Apple Watch track vital metrics such as heart rate, blood oxygen levels, and sleep patterns, enabling users to manage their health proactively. For instance, IoT wearables have alerted users to early signs of health issues, prompting timely medical intervention and improving outcomes. The global wearable technology market is projected to reach $104.39 billion by 2027, underscoring its growing influence in healthcare.[38]

Integration into daily life and business operations highlights its potential to drive efficiency and innovation on a global scale.

The Internet of Things plays a critical role in optimizing supply chains and logistics. Real-time tracking devices embedded in shipping containers monitor location, temperature, and humidity, ensuring goods arrive in optimal condition. Companies like DHL leverage IoT to enhance supply chain visibility, reduce delays, and improve efficiency. IoT adoption in logistics is expected to grow significantly, with the market forecast to reach $100 billion by 2030.[39] Devices like thermostats, lighting systems, and voice assistants integrate seamlessly in smart homes to improve convenience and energy efficiency. Products like Google Nest and Amazon Echo provide real-time automation and control, offering users a more connected and intuitive living experience. According to Statista, smart home devices are expected to exceed 1.8 billion units globally by 2025.[40]

IoT enables smart factories in manufacturing where machines communicate autonomously to optimize production. Predictive maintenance systems, powered by IoT sensors, monitor equipment health and reduce downtime by addressing issues before failures occur. Industries implementing IoT solutions report improved productivity and cost savings, driving its adoption across the sector.[41]

This ecosystem continues to expand, connecting devices and data in ways that redefine workflows, enhance user experiences, and enable more informed decision-making across industries. Its integration into daily life and business operations highlights its potential to drive efficiency and innovation on a global scale.

5G Connectivity

The global rollout of 5G networks is transforming digital interactions by delivering faster, more reliable connections and ultra-low latency. This connectivity is critical for enabling advanced mobile-first experiences, supporting the Internet of

Things, and powering immersive technologies like augmented reality and virtual reality). With speeds up to 100 times faster than 4G and significantly reduced latency, 5G technology redefines what's possible in user experience.

For mobile-first businesses, 5G provides the bandwidth to handle complex applications seamlessly. Streaming services, for example, can deliver ultra-high-definition content without buffering, while mobile apps can load and function more quickly, improving overall user satisfaction. 5G's capacity to connect a massive number of devices simultaneously is a game-changer for IoT ecosystems. Smart homes, autonomous vehicles, and connected healthcare devices benefit from the network's ability to facilitate real-time communication. According to a report by Ericsson, by 2028, over 5 billion 5G subscriptions are expected worldwide, driving significant growth in IoT adoption.[42]

In AR and VR, 5G's low latency and high bandwidth are critical. These applications require near-instantaneous data transfer to function effectively. For example, stadiums equipped with 5G networks can offer fans AR-driven experiences, such as viewing real-time stats, instant replays, and interactive highlights through their mobile devices. This integration enhances the in-person sports experience, making it more engaging and personalized.

Beyond entertainment, industries like healthcare and manufacturing leverage 5G for innovative use cases. In telemedicine, 5G allows for real-time consultations and remote surgeries with minimal latency. In manufacturing, it supports smart factories where machines communicate and adjust operations dynamically.[43] The deployment of 5G is not just an upgrade in connectivity. It's a foundational technology enabling the next wave of digital transformation across industries. Its role in driving faster, smarter, and more interactive experiences underscores its significance in the future of UX and CX.

Why UX Is Essential to Emerging Technologies

Adopting cutting-edge technologies without user-centered design risks underutilization and poor adoption. UX ensures these innovations are accessible, intuitive, and aligned with user needs. As businesses navigate the next wave of digital transformation, UX will be the key to unlocking the full potential of emerging technologies. Delivering value for both users and organizations.

Some Cautionary Tales

In today's rapidly evolving tech landscape, businesses are eager to adopt cutting-edge technologies like AI, mixed-reality devices, and automation systems to gain a competitive edge. However, this race to innovate often leads to a critical oversight: the user experience. Neglecting UX while implementing new technologies can result in poor adoption rates, user frustration, reputational damage, and even product failure.

One of the most prominent cautionary tales is Juicero, a high-tech juicing startup that exemplified the risks of prioritizing innovation over user value. The company launched a $400 juicing machine requiring proprietary juice packs and a Wi-Fi connection. Customers quickly discovered they could squeeze the packs by hand, rendering the machine unnecessary. This revelation triggered widespread ridicule, with critics dubbing the device emblematic of Silicon Valley excess. Juicero shut down in 2017, a failure that underscores the importance of designing technology that enhances the user experience rather than complicates it.[44]

> *This race to innovate often leads to a critical oversight: the user experience.*

Another striking example is Microsoft's AI chatbot Tay, launched in 2016 to engage with users on Twitter. Tay was designed to learn from conversations in real time. However, within 24 hours of its debut, users manipulated the chatbot, exposing it to harmful interactions that led it to generate offensive and inappropriate content.[45] This public debacle highlighted the risks of deploying AI without robust safeguards or UX considerations. Tay's failure demonstrates that even advanced technologies can backfire without thoughtful planning and management of user interactions.

The debut of Apple's Vision Pro, a $3,499 mixed-reality headset, offers a more nuanced example. The product showcases Apple's engineering prowess and ambition to define the future of spatial computing. However, its high price point and limited practicality for everyday users have sparked criticism.[46] While the Vision Pro pushes technological boundaries, its reception raises questions about balancing innovation with usability and accessibility. Apple's approach emphasizes the need to align groundbreaking features with real-world user needs to ensure broader adoption.

These examples demonstrate the critical importance of UX in adopting new technologies. Companies must balance innovation and usability to avoid alienating users or undermining their ambitions. By placing user needs at the center of technological advancements, businesses can ensure their innovations deliver meaningful value, foster adoption, and build lasting trust.

Ethics in AI

As businesses increasingly rely on artificial intelligence, ethical design becomes critical. AI has immense potential to enhance user experiences, but without careful oversight, it can perpetuate bias, compromise privacy, or erode trust. UX teams are vital in ensuring AI systems are designed and deployed responsibly.

Respecting User Privacy

Key principles of ethical AI design emphasize the importance of building transparent, fair, and respectful systems of user privacy. Transparency involves communicating when users interact with AI rather than humans, ensuring clarity and trust. For example, after much controversy, Google's AI assistant Duplex now explicitly identifies itself as AI during phone interactions, making users aware of its nature.[47] Bias mitigation focuses on identifying and addressing biases in AI algorithms to prevent discriminatory outcomes. Companies like IBM have developed tools to audit AI models for fairness and accountability, setting ethical implementation standards.[48]

Privacy by design ensures that data collection is transparent, consensual, and limited to its intended purpose. This includes providing clear opt-in mechanisms and explaining how data will be used, safeguarding user trust and data security. These principles are essential for creating AI systems that are ethical, reliable, and aligned with user needs.

Cases like Amazon's recruitment AI, which was discontinued after it was found to favor male applicants for technical roles, underscore the importance of ethical AI.[49] This incident highlights the need for rigorous testing and oversight in AI development to prevent harm and ensure equity.

The Risk of AI Adoption Without UX

The rapid adoption of AI highlights the risks of embracing transformative technology without prioritizing UX. Many organizations focus on integrating AI into their products and services without fully considering its impact on usability. This oversight often results in tools that feel alienating, overly complex, or untrustworthy. For instance, companies that deploy AI-driven customer support systems without refining the user interface may inadvertently frustrate customers who struggle to navigate impersonal or rigid systems. Such outcomes not only diminish the intended benefits of AI but also erode user trust and satisfaction.

To mitigate these risks, businesses must embed UX considerations into every stage of technology development. A thoughtful approach includes several key strategies:

Conduct Thorough User Research; Before introducing new tools, it is essential to understand the needs, pain points, and preferences of the target audience. This ensures that the technology addresses real user problems and aligns with expectations.

Prioritize Usability in Design: Intuitive, accessible, seamless systems are critical for enhancing user experiences. Employing iterative testing and prototyping during development helps uncover usability issues early, allowing teams to refine solutions before launch.

Implement Robust Safeguards: For AI, systems must include mechanisms to monitor and prevent unintended behaviors. Safeguards ensure ethical implementation and foster trust by making interactions feel secure and reliable.

Test for Alignment with Real-World Needs: Balancing innovation with practicality is vital to delivering meaningful value. Businesses should validate that their technology enhances everyday user scenarios, ensuring it integrates seamlessly into their lives.

By addressing these considerations, businesses can avoid common pitfalls and maximize the potential of AI and other emerging technologies.

The UX Imperative for Technological Success

In the pursuit of technological advancement, businesses must prioritize UX to ensure success. As we've discussed, high-profile failures, such as Juicero's over-engineered juicing machine and Microsoft's AI chatbot Tay, demonstrate that even the most advanced technologies can falter without a user-centered approach.

These cases underscore the importance of embedding UX principles into developing and implementing new technology. By focusing on how technology serves people, businesses can achieve higher adoption rates, enhanced user satisfaction, and long-term success. In the era of AI and rapid innovation, companies that prioritize user-centered design will thrive.

Sephora provides a compelling example of how a user-centered approach to emerging technologies can drive innovation and business growth. By strategically integrating AI, AR, and voice interfaces, Sephora has transformed the beauty retail experience, seamlessly merging the convenience of online shopping with the immersive qualities of in-store interactions. Sephora uses AI to analyze customer data, including browsing behavior and purchase history, to deliver tailored product recommendations. This personalized approach enhances engagement and loyalty, creating a shopping experience uniquely tailored to each individual.

In the era of AI and rapid innovation, companies that prioritize user-centered design will thrive.

The Sephora Virtual Artist app, developed with AR pioneer ModiFace, allows customers to try on makeup products virtually.[50] Since its launch, users have tried on over 200 million shades, with over 8.5 million visits to the feature. This innovative tool reduces purchase hesitation, increasing conversions and greater customer confidence.

Sephora's AI-powered chatbots, such as the Sephora Assistant, simplify customer interactions by enabling tasks like booking makeover appointments. These tools enhance the customer experience and streamline operations, reducing administrative costs and improving efficiency. Sephora's commitment to digital innovation has driven remarkable financial growth. Its e-commerce net sales increased from approximately $580 million in 2016 to over $3 billion in 2023, representing a significant rise in online revenue.[51]

By leveraging emerging technologies with a focus on UX, Sephora has redefined the beauty retail experience. Its innovative tools meet customer expectations and set new industry standards, ensuring sustained growth and solidifying its position as a leader in its market.

The Business Case for Accessibility

Accessibility is no longer optional. It is essential to creating digital experiences that effectively serve all users. Inclusive design ensures usability for diverse audiences, including people with disabilities, older adults, and those relying on assistive technologies. By embedding accessibility into the design process, businesses demonstrate a commitment to equity and inclusion, expand their reach, enhance user satisfaction, and strengthen their brand.

Effective accessibility design incorporates key features to meet the diverse needs of users with varying abilities. Screen reader compatibility ensures that interactive elements like buttons, forms, and menus are labeled correctly. Hence, they work seamlessly with screen readers, allowing visually impaired users to navigate digital interfaces effectively.

Adjustable text sizes enable users to resize text for improved readability without compromising functionality or the integrity of the layout. Keyboard navigation is critical for users who cannot use a mouse, allowing them to interact with digital content entirely through keyboard inputs.

Additionally, voice commands provide hands-free interaction for individuals with motor impairments, making digital tools accessible through spoken instructions. These features create a more inclusive and user-friendly experience for all individuals.

The Impact of Accessibility

Companies that prioritize accessibility often stand out as leaders in innovation and social responsibility. Microsoft, for example, has introduced transformative accessibility features, including Eye Control in Windows.[52] This functionality enables users to navigate their computers using only eye movements, breaking barriers for individuals with motor impairments. Such innovations enhance inclusivity and reinforce Microsoft's reputation as a socially conscious brand.

Inclusive Design

Accessibility isn't just about doing the right thing. It's also good for business. Inclusive design broadens a company's potential audience, improves user satisfaction, and reduces the risk of legal and financial repercussions. The rise in digital accessibility lawsuits highlights the importance of compliance. By prioritizing accessibility, businesses align with ethical standards and future-proof their digital platforms. This strategy ensures usability for all users while mitigating legal risks. Inclusive design is a win-win strategy that enhances the user experience and demonstrates a commitment to equity and innovation.

Incorporating accessibility and ethical design principles is both a moral imperative and a strategic business advantage. Accessible and ethical designs expand market reach by making products usable for everyone, including the estimated 1 billion people globally who experience some form of disability, according to the World Bank. This inclusivity opens businesses to more extensive and diverse audiences. Building trust through transparent and inclusive practices also fosters long-term loyalty and brand advocacy, as users are more likely to support companies that align with their values. Finally, mitigating risks by proactively addressing accessibility and ethical concerns reduces the likelihood of legal challenges, reputational damage, and customer attrition, positioning businesses as responsible and forward-thinking in a competitive marketplace.

Increase in Litigation

In recent years, there has been a notable increase in litigation against corporations for failing to address digital accessibility. In 2023 alone, over 4,000 digital accessibility lawsuits were filed, marking a significant rise from previous years.[53] This upward trend underscores the growing legal risks for businesses that neglect to make their digital platforms accessible to individuals with disabilities. Most of these lawsuits have been concentrated in states like New York and Florida, where legal precedents and state laws are particularly favorable to plaintiffs in accessibility cases.

This surge in litigation highlights the imperative for companies to proactively ensure their digital content complies with accessibility standards to mitigate legal exposure and promote inclusivity.

Embedding Ethics and Accessibility into UX Processes

Embedding ethical design and accessibility into UX workflows ensures businesses create inclusive, transparent, and trustworthy experiences. This involves conducting usability testing with diverse user groups, including people with disabilities, to identify and address barriers. It also means implementing design systems prioritizing accessibility from the outset, ensuring that inclusivity is not an afterthought but a foundational principle. Regular audits of AI systems are essential to maintain fairness, accuracy, and privacy compliance, ensuring that technological advancements align with ethical standards.

By embedding these practices into their processes, companies enhance their user experience while positioning themselves as leaders in responsible innovation. This commitment to accessibility and ethics builds trust, fosters customer loyalty, and ensures that products serve all users equitably. Businesses that adopt these principles demonstrate a dedication to doing what is right and a strategic understanding of the benefits of inclusivity and fairness.

Preparing for the Future with UX

In a rapidly evolving world, UX equips businesses to navigate change, adapt to new challenges, and confidently embrace innovation. Emerging technologies like AI, AR, voice interfaces among others, hold immense potential to transform industries. However, their success depends on one critical factor: how well they meet user needs.

Turning Ideas into Value

Innovation often starts with bold ideas, but without user feedback, even the most advanced technologies can fail to deliver value. UX bridges this gap by ensuring new tools and features address real problems and enhance user experiences. Businesses can use techniques like rapid prototyping and usability testing to validate ideas early, enabling teams to refine concepts before scaling.

For instance, a retail company experimenting with an AI-driven chatbot might develop a prototype to handle customer inquiries. By testing it with a small user group, the team can identify pain points, such as unclear responses or limited functionality, and make improvements before a full rollout. This iterative approach ensures the chatbot provides real value, improves customer satisfaction, and aligns with business goals.

Fostering Collaboration Across Teams

Innovation rarely happens in isolation. It requires collaboration across IT, marketing, operations, and product development functions. UX fosters this alignment by creating a shared focus on user outcomes, ensuring that new technologies are seamlessly integrated into the broader business strategy.

For example, an e-commerce platform introducing AR features to allow virtual try-ons might require input from product teams for design, IT for technical feasibility, and marketing to promote the feature. A user-centered approach ensures all teams align around the shared goal of enhancing the customer experience, avoiding silos that can derail projects or reduce their impact.

Scalable Experimentation

UX principles enable businesses to experiment with emerging technologies in a scalable way. By starting small and gathering user feedback, companies can identify what works, what doesn't, and where adjustments are needed. This approach reduces the risks associated with large-scale implementations while accelerating time-to-value. For instance, when Spotify began testing its personalized Discover Weekly playlists, it used a controlled rollout to gather insights on user preferences and engagement.[54] By iterating based on this feedback, Spotify fine-tuned the feature into one of its most successful offerings, driving increased retention and user satisfaction.

Positioning Your Business to Lead

At its core, UX is a strategic enabler that ensures innovation focuses on solving real user problems. While emerging technologies like AI, AR/VR, and IoT may dominate the headlines, the true differentiator lies in how these technologies are applied to meet user needs effectively and intuitively. UX ensures that experimentation with these innovations isn't just a race to adopt the latest tools but a deliberate effort to create meaningful outcomes that enhance customer and employee experiences.

Prioritizing UX transforms cutting-edge technology into a competitive advantage. For example, integrating AI into customer support can reduce wait times and increase satisfaction, provided the system is designed with the user in mind, offering intuitive interfaces and clear communication. Similarly, AR/VR technologies can redefine shopping or training experiences only when accessible and aligned with real-world needs. UX practitioners are critical in bridging this gap, turning potential into performance by ensuring these solutions work seamlessly for users.

Organizations that embed UX into their innovation strategies are better equipped to adapt to change and thrive in dynamic markets. They foster trust and loyalty by consistently delivering value through thoughtful, user-centered design. This positions them not just as participants in the digital age but as leaders shaping its future.

In the final chapter, we'll explore how companies across industries implement UX and demonstrate how it drives growth, strengthens relationships, and delivers measurable results. These real-world examples will bring to life the transformative power of UX and provide a roadmap for unlocking its full potential in your organization.

NOTES

NOTES

Chapter 10: UX in Action

Good design is the most important way to differentiate ourselves from our competitors.

Yun Jong Yong
CEO—Samsung Electronics

Throughout this book, we've explored how UX can transform businesses: aligning with strategy, enhancing customer experiences, empowering employees, and driving measurable outcomes. But UX is not just a methodology or a toolkit; it's a way to create value by solving real user problems. It bridges the gap between vision and execution, ensuring innovation is impactful and sustainable. At the heart of this transformation is ROI. Real, quantifiable results that show the value of investing in user-centered design. We've seen companies achieve returns exceeding 9,900% by embedding UX into their strategy, aligning with business goals, and scaling thoughtfully. The frameworks, examples, and strategies in this book have provided you with everything you need to achieve similar success within your organization.

This final chapter is about action. It's about taking the lessons learned and applying them in ways that drive immediate and long-term value. Whether you're starting small or scaling a mature UX practice, the principles in this book position you to lead with UX and unlock its full potential for your business.

UX as a Strategic Input

One of the core takeaways is that UX is not just about making things look better. It's about making them work better for both users and the business. By aligning UX initiatives with measurable business goals, such as increasing revenue, reducing costs, and improving customer retention, you elevate UX from a design function to a strategic driver. Companies prioritizing UX aren't just refining touchpoints; they're creating systems that enable smarter decisions, better experiences, and sustained growth. As discussed, businesses like IBM, Delta Airlines, and Capital One have leveraged UX to reduce inefficiencies, boost satisfaction, and create competitive differentiation. These aren't just theoretical examples. They're proof that UX delivers when aligned with business objectives.

Alignment and Integration

The power of UX lies in its ability to unify efforts across teams and functions. When UX is integrated into product development, marketing, customer support, and internal operations, it becomes a shared language for solving problems and driving results. Throughout this book, we've emphasized the importance of collaboration. From journey mapping exercises that identify cross-functional pain points to aligning UX metrics with key performance indicators (KPIs), the tools and strategies we've outlined ensure that UX efforts contribute directly to your organization's success.

Starting Small and Scaling

For organizations just beginning their UX journey, the best approach is often to start small. Target a specific area where UX can deliver immediate results, such as streamlining a customer onboarding process or redesigning an internal tool to save employee time. These early wins build momentum and demonstrate value, creating a foundation for scaling UX across the organization. As your UX practice grows, so does its impact. By expanding UX principles into other areas, such as customer support, operational workflows, or emerging technology, you position your business for sustainable growth. As you scale, investing in UX leadership and a strong community of practice ensures your efforts remain aligned and impactful.

With these lessons in mind, you can begin or expand your UX initiatives. The ROI of UX is within reach, and the steps outlined in this book will help you achieve it. Let's explore how to bridge the gap between vision and execution, ensuring that UX becomes a lasting source of value for your business.

Bridging Vision and Execution

Achieving the ROI of UX isn't just about knowing the right principles. It's about consistently applying them in ways that solve real problems. This requires translating high-level strategies into actionable initiatives that resonate with your team and deliver results for your customers. Here's how to make that connection, step by step:

Identify High-Impact Opportunities

The first step is to focus on areas where UX can deliver immediate and measurable value. These might include simplifying a critical workflow, improving a frustrating customer touchpoint, or addressing inefficiencies in an internal system. Start by leveraging user feedback and analytics to pinpoint friction points, then prioritize those with the greatest potential to impact business goals. For example, if customer retention is a challenge, investigate the onboarding process. Are users finding it intuitive? Are they reaching moments of value quickly? Identifying and addressing specific issues through user-centered design can directly influence metrics like churn rate and customer lifetime value.

Set Clear, Measurable Goals

Every UX initiative should be tied to a measurable outcome. Metrics like task completion rates, Net Promoter Scores (NPS), or employee satisfaction scores serve as leading indicators of business performance. These metrics don't just show progress. They provide a way to demonstrate ROI to leadership. Consider the example of redesigning an internal dashboard for a sales team. The goal might be to reduce task completion time by 20% while increasing employee adoption rates. Tracking these metrics before and after the redesign creates a clear narrative about how UX contributes to operational efficiency and employee engagement.

Collaborate Across Teams

Successful UX initiatives rarely happen in isolation. To ensure alignment and support, they require collaboration across product management, engineering, marketing, and operations. This cross-functional approach not only improves the quality of the solution but also reinforces the value of UX as a unifying force within the organization. Encourage open communication and shared accountability. For example, product managers can be involved in usability testing to ground feature prioritization in real user insights. Similarly, work with marketing to ensure customer-facing interfaces align with the brand voice and message. These partnerships help ensure that UX is embedded into every stage of the business process.

Prototype, Test, and Iterate

The essence of UX is iterative problem-solving. Rather than perfecting a solution on the first attempt, focus on building prototypes, testing them with users, and refining them based on feedback. This approach minimizes risk while maximizing impact, ensuring solutions are effective and aligned with user needs. For instance, before rolling out a redesigned feature, test it with a small segment of users. Gather qualitative and quantitative feedback, analyze the results, and make adjustments. When the solution is scaled organization-wide, it's been validated and optimized for success.

Communicate Success Widely

Once a UX initiative delivers results, share those outcomes broadly within your organization. Use data and user stories to illustrate the impact, whether it's increased efficiency, higher satisfaction, or improved financial performance. This transparency builds buy-in for future projects and reinforces the importance of UX as a strategic function. By bridging vision and execution through these practical steps, you create a system where UX consistently drives measurable value. It's not about one-off wins. It's about building momentum for lasting transformation.

Sustaining Momentum

Driving ROI through UX requires more than isolated successes. It demands a sustained commitment to user-centered thinking across every level of your organization. To ensure that your UX efforts continue to deliver value, you must embed them into your culture, processes, and long-term strategy.

Make UX a Core Business Capability

For UX to thrive, it must be seen as a core business capability, not just a set of tactical tools or a design department function. This means elevating UX to the same level of strategic importance as marketing, product management, or operations. Leaders should ensure that UX practitioners are involved in key business decisions and treated as equal partners in defining and delivering outcomes. One way to institutionalize UX is by aligning it with your organization's strategic goals. For example, if your business aims to reduce operational costs, ensure that UX teams are focused on redesigning internal tools or processes that impact efficiency. When UX is directly tied to organizational objectives, it becomes an integral part of your business.

Foster a Culture of Continuous Improvement

Sustaining the momentum of UX requires an environment where learning, iteration, and adaptation are encouraged. A culture of continuous improvement ensures that UX efforts don't plateau but instead evolve in response to changing user needs and market conditions. Establish systems for gathering ongoing user feedback, such as surveys, usability testing, or analytics, to identify opportunities for refinement and ensure that each iteration delivers incremental value. Acknowledge and share successes at every stage of a UX initiative, from early prototypes to post-launch improvements, to keep teams motivated and reinforce the value of their work. Equip your UX teams with tools, training, and resources to stay ahead of emerging trends and technologies, enabling them to bring fresh ideas and innovative solutions that drive your business forward.

Scale UX Thoughtfully

As your organization grows, scaling UX capabilities must be approached with intention and planning. Rapid growth without a clear structure can dilute the impact of UX and lead to inconsistent results. Instead, align UX expansion efforts with your company's evolving needs. Begin by appointing experienced UX leaders who can advocate for user-centered design at the executive level and ensure alignment with strategic priorities. Establish standardized processes, such as design systems or research protocols, to create efficiency and maintain quality as teams expand their efforts. Strengthen connections with other departments, including HR, operations, and customer service, to extend the reach and influence of UX, amplifying its impact across the organization. Thoughtful scaling ensures that UX continues to deliver meaningful value as your business evolves.

Future-Proof Your UX Strategy

The rapid pace of technological change demands that businesses stay ahead of emerging trends to remain competitive. Sustaining UX momentum requires a forward-thinking approach that anticipates and adapts to these shifts. Stay informed about advancements like AI, AR, and voice interfaces, and thoughtfully integrate these technologies to enhance user experiences while aligning them with user needs and preferences. Monitor market shifts, including evolving user expectations, industry standards, and competitive landscapes, to ensure your UX strategy remains relevant and impactful. Commit to accessibility and ethical design as these areas grow in importance, building trust and loyalty among users. By prioritizing these elements, businesses position themselves to lead in a dynamic and ever-evolving marketplace.

Building the Foundation for Long-Term Success

Sustaining momentum in UX isn't about maintaining the status quo. It's about fostering a culture where user-centered thinking drives continuous improvement, innovation, and measurable impact. By embedding UX into your organization's DNA, you create a foundation for growth that adapts to new challenges and opportunities. The tools, insights, and strategies outlined in this book have given you the framework to unlock the enterprise value of UX. In the next section, we'll see how these principles translate into action, examining real-world examples of businesses that have successfully harnessed UX to transform their organizations and achieve extraordinary results.

Unlocking the Full Potential of UX

The journey through this book has illuminated one central truth: UX is not just a design discipline; it's a strategic asset that drives measurable outcomes across every facet of your business. From customer loyalty and operational efficiency to innovation and employee engagement, user-centered design principles create ripple effects that enhance your organization's performance and resilience.

We've explored UX's ROI, alignment with business strategy, connection to customer and employee experiences, and the practical steps needed to scale its influence across your organization. At every stage, the message is clear: Businesses that prioritize UX unlock opportunities to deliver value in ways that resonate deeply with users and the bottom line.

Key Takeaways for Your UX Journey

The ROI of UX Is Real: From increased revenue to cost savings and higher retention rates, the financial benefits of UX are undeniable. By aligning UX metrics with business KPIs, you ensure that every user interaction contributes directly to your strategic goals.

UX Aligns Strategy with Execution: UX bridges the gap between high-level business objectives and the experiences your users and employees encounter daily. It transforms vision into reality by aligning processes, systems, and tools with user needs.

Scaling UX Requires Commitment: Expanding UX's influence across your organization involves thoughtful planning, clear goals, and dedicated leadership. A thriving UX practice isn't built overnight. It's cultivated through sustained effort and a focus on collaboration, consistency, and measurable outcomes.

Emerging Technologies Are an Opportunity: From AI and AR to blockchain and 5G, the rapid evolution of technology creates new possibilities for UX innovation. By thoughtfully integrating these tools, you can stay ahead of the curve and redefine your business's possibilities.

Ethics and Accessibility Are Non-Negotiable: As digital landscapes evolve, prioritizing ethical design and accessibility ensures your business meets today's users' expectations while building trust and fostering inclusivity.

The insights and strategies in this book are only as powerful as the action they inspire. Whether you're just beginning to explore UX or scaling its impact across your organization, now is the time to take the next step. Start with small, high-impact projects that demonstrate value, align UX efforts with measurable outcomes, and cultivate a culture of continuous improvement. Success in today's market isn't just about creating great products but delivering exceptional experiences that build lasting relationships. UX gives you the framework to do just that, helping you navigate change, seize opportunities, and differentiate your business in a competitive world.

The Path Forward

The future belongs to businesses that embrace UX as a driver of growth, innovation, and resilience. By embedding UX into your strategy, fostering collaboration across teams, and aligning with the evolving needs of your users, you're positioning your organization to lead in an increasingly dynamic marketplace. UX is a mindset. It's about seeing your business through your users' eyes and making decisions prioritizing their success. When you invest in UX, you're not just improving interfaces or workflows but building the foundation for long-term value and impact.

The opportunity is here, and the tools are in your hands. It's time to lead your organization into the future. One that is better by design.

About the Author

Mike Kuechenmeister is a UX designer and consultant with nearly three decades of experience, including leadership roles at Fortune 100 companies like Northwestern Mutual and Optum Health. As the founder of Supergreen, he leads a team of experts in UX strategy, design, research and content, driving growth for companies commited to doing good.

Learn more at www.supergreen.us.

Endnotes

All links effective as of publication date. For easier reference, please visit www.supergreen.us/betterbydesign

1 The Total Economic Impact™ Of IBM's Design Thinking Practice
https://www.ibm.com/design/thinking/static/Enterprise-Design-Thinking-Report-8ab1e9e1622899654844a5fe1d760ed5.pdf

2 The Six Steps For Justifying Better UX
https://www.forrester.com/report/The-Six-Steps-For-Justifying-Better-UX/RES117708

3 The business value of design
https://www.mckinsey.com/capabilities/mckinsey-design/our-insights/the-business-value-of-design?cid=soc-web

4 BofA Unifies Mobile Apps for Banking, Investing, and Retirement Into One Personalized Digital Experience
https://newsroom.bankofamerica.com/content/newsroom/press-releases/2024/03/bofa-unifies-mobile-apps-for-banking--investing--and-retirement-.html

5 Insights From eCommerce CRO Apex Predators: Booking.com
https://vwo.com/blog/cro-best-practices-booking/

6 13 Checkout Optimization Tips To Increase Ecommerce Revenue (2024)
https://www.shopify.com/blog/checkout-process-optimization

7 Delta upgrades app to Fly Delta 6.0 with new and updated features, functionality and experiences
https://www.futuretravelexperience.com/2024/05/delta-upgrades-app-to-fly-delta-6-0-with-new-and-updated-features-functionality-and-experiences/

8 Stepping into a new world of workforce management. Case study: Kaiser Permanente
https://www.myamericannurse.com/stepping-into-a-new-world-of-workforce-management-technology-can-help-you-make-the-right-staffing-choices/

9 Faster, Smoother, Better: How Express Checkout Boosts Conversions and Revenue
https://blog.2checkout.com/how-express-checkout-boosts-conversions/

10 Conversion Optimization UX: 10 Best Practices
https://landingi.com/conversion-optimization/ux/#:~:text=According%20to%20statistics%20reported%20by,up%20to%20a%20staggering%20400%25

11 Reducing Development Costs with Effective UX/UI
https://piximind.com/en/blog/cost/reducing-development-costs-effective-uxui

12 A Tech Exec's Guide To Workflows
https://www.forrester.com/report/a-tech-execs-guide-to-workflows/RES177781

13 Bineo UX Case Study: Designing the First 100% Digital Bank in Mexico
https://www.theuxda.com/blog/bineo-ux-case-study-designing-first-100-digital-bank-in-mexico

14 Perfect Strangers: How Airbnb is building trust between hosts and guests
https://news.airbnb.com/perfect-strangers-how-airbnb-is-building-trust-between-hosts-and-guests/

15 Budget Planning Guide 2025: Customer Experience
https://www.forrester.com/report/budget-planning-guide-2025-customer-experience/RES181174

16 The business value of design
https://www.mckinsey.com/capabilities/mckinsey-design/our-insights/the-business-value-of-design

17 Delta upgrades app to Fly Delta 6.0 with new and updated features, functionality and experiences
https://www.futuretravelexperience.com/2024/05/delta-upgrades-app-to-fly-delta-6-0-with-new-and-updated-features-functionality-and-experiences/

18 Unwavering Commitment to Operational Excellence for You and Your Travelers
https://pro.delta.com/content/agency/us/en/news/news-archive/2023/november-2023/unwavering-commitment-to-operational-excellence-for-you-and-your.html

19 SEC Charges Knight Capital With Violations of Market Access Rule
https://www.sec.gov/newsroom/press-releases/2013-222

20 Target deepens bets on supply chain, inventory management
https://www.supplychaindive.com/news/Target-supply-chain-inventory-management-Brian-Cornell-Q4/709914/

21 Stepping into a new world of workforce management. Case study: Kaiser Permanente
https://www.myamericannurse.com/stepping-into-a-new-world-of-workforce-management-technology-can-help-you-make-the-right-staffing-choices/

22 Medication Errors for Admitted Patients Drop When Pharmacy Staff Take Drug Histories in ER
https://www.cedars-sinai.org/newsroom/medication-errors-for-admitted-patients-drop-when-pharmacy-staff-take-drug-histories-in-er/

23 Why Delta focuses on change management in CX technology rollouts
https://www.customerexperiencedive.com/news/delta-change-management-cx-technology-agents-ccw/724288/

24 Improving project management processes
https://www.pmi.org/learning/library/improving-accelerated-process-improvement-intel-7023

25 Want Happy Customers? Learn from the Southwest Airlines® Employee-First Mindset.
https://www.salesforce.com/resources/customer-stories/southwest-airlines-employee-experience/

26 Design Firm Adaptive Path Acquired By Capital One
https://techcrunch.com/2014/10/02/adaptive-path-acquired-by-capital-one/

27 Leading banking apps in the United States in 2023, by downloads
https://www.statista.com/statistics/1381325/us-leading-banking-apps-by-downloads/

28 Capital One Mobile app analytics for March 21
https://www.similarweb.com/app/apple/407558537/#overview

29 Silicon Valley Product Group https://www.svpg.com/

30 Spotlight on 2024 Gartner Hype Cycle™ for Emerging Technologies
https://www.gartner.com/en/articles/hype-cycle-for-emerging-technologies

31 The Total Economic Impact™ Of IBM's Design Thinking Practice
https://www.ibm.com/design/thinking/static/Enterprise-Design-Thinking-Report-8ab1e9e1622899654844a5fe1d760ed5.pdf

32 Gen AI Use Cases in Telecom Helping Operators Optimize and Innovate
https://www.processica.com/articles/gen-ai-use-cases-in-telecom-helping-operators-optimize-and-innovate/

33 20+ Telling Personalization Statistics [2023]: Investment, ROI, and More
https://www.zippia.com/advice/personalization-statistics/

34 The State Of Consumer Personalization, 2023
https://www.forrester.com/report/the-state-of-consumer-
personalization-2023/RES180120

**35 The Role of Augmented Reality (AR) in Retail: A Case Study on
IKEA**
https://footar.co/the-role-of-augmented-reality-ar-in-retail-a-case-study-
on-ikea/

**36 VR and AR in Training: Immersive Learning for the Modern
Workforce**
https://www.ignitehcm.com/blog/vr-and-ar-in-training-immersive-learning-
for-the-modern-workforce

**37 Blockchain in the food supply chain - What does the future look
like?**
https://tech.walmart.com/content/walmart-global-tech/en_us/blog/post/
blockchain-in-the-food-supply-chain.html

**38 Wearable Technology Market Size, Share & Trends Analysis
Report**
https://www.grandviewresearch.com/industry-analysis/wearable-
technology-market

**39 IoT in Warehouse Management Market Size, Share & COVID-19
Impact Analysis**
https://www.fortunebusinessinsights.com/iot-in-warehouse-management-
market-107383

**40 Internet of Things (IoT) and non-IoT active device connections
worldwide from 2010 to 2025**
https://www.statista.com/statistics/1101442/iot-number-of-connected-
devices-worldwide/

**41 IoT value set to accelerate through 2030: Where and how to
capture it**
https://www.mckinsey.com/capabilities/mckinsey-digital/our-insights/iot-
value-set-to-accelerate-through-2030-where-and-how-to-capture-it

**42 Ericsson Mobility Report: Global 5G growth amid macroeconomic
challenges**
https://www.ericsson.com/en/press-releases/2022/11/ericsson-mobility-
report-global-5g-growth-amid-macroeconomic-challenges

43 Everything you need to know about 5G
https://www.qualcomm.com/5g/what-is-5g#:~:text=5G%20is%20
designed%20to%20not,and%20connecting%20the%20massive%20IoT.

**44 Juicero, maker of the doomed $400 internet-connected juicer, is
shutting down**
https://www.theverge.com/2017/9/1/16243356/juicero-shut-down-lay-off-
refund

45 Microsoft chatbot is taught to swear on Twitter
https://www.bbc.com/news/technology-35890188

46 Review: Apple Vision Pro
https://www.wired.com/review/apple-vision-pro/

47 Google's 'deceitful' AI assistant to identify itself as a robot during calls
https://www.theguardian.com/technology/2018/may/11/google-duplex-ai-identify-itself-as-robot-during-calls

48 What is AI governance?
https://www.ibm.com/think/topics/ai-governance

49 Insight - Amazon scraps secret AI recruiting tool that showed bias against women
https://www.reuters.com/article/world/insight-amazon-scraps-secret-ai-recruiting-tool-that-showed-bias-against-women-idUSKCN1MK0AG/

50 Sephora Virtual Artist Adds Virtual Try On Of Thousands Of Eyeshadow Shades
https://www.multivu.com/players/English/7926153-sephora-virtualartist-app/

51 E-commerce net sales of sephora.com from 2016 to 2025
https://www.statista.com/forecasts/1383797/sephora-revenue-development-ecommercedb

52 Eye Control for Windows 10
https://www.microsoft.com/en-us/garage/wall-of-fame/eye-control-windows-10/#:~:text=Eye%20Control%20makes%20Windows%20 10,experience%20using%20only%20their%20eyes.

53 How Companies Should Mitigate the Risks of Digital Accessibility Litigation
https://todaysgeneralcounsel.com/how-companies-should-mitigate-the-risks-of-digital-accessibility-litigation/

54 Here's the Story Behind Spotify's Coolest Feature
https://time.com/4131520/spotify-discover-weekly-playlists/

www.ingramcontent.com/pod-product-compliance
Lightning Source LLC
Chambersburg PA
CBHW071551200326
41519CB00021BB/6694